CW00751443

The Collectors

Creating Hans Sloane's extraordinary herbarium

Edited by Mark Carine

Published by the Natural History Museum, London

First published by the Natural History Museum,
Cromwell Road, London SW7 5BD

ISBN 9780565094881

A catalogue record for this book is available
from the British Library

10 9 8 7 6 5 4 3 2 1

Copy-edited by Celia Coyne
Designed by Bobby Birchall, Bobby&Co.
Reproduction by Saxon Digital Services, Norfolk
Printed by Toppan Leefung Printing Limited, China

Front cover Ornamental flowers from the
herbarium of George London.
Back cover Parrot tulips from Robert Uvedale's
collection.

CONTENTS

FOREWORD

TIM LITTLEWOOD

Requiring a blend of inquisitiveness, tenacity, thoughtfulness and a certain amount of indifference to potential danger, the journey towards successful exploration and notable discovery is borne of a particular mindset. It is not uncommon for us to think of explorers and discoverers as passionate pioneers and fearless adventurers, but the reality is often at odds with this. Often it is the stoic, stubborn, and single-minded individual's perseverance which leads to revelations that leave us in awe of the natural world. Exploration and discovery are key pillars of science, yet they are innately human endeavours.

Some years ago, I was lucky enough to visit the public library in Killyleagh, birthplace of Sir Hans Sloane, to see a poster on the wall of Hans as a boy. Stooped with stick and net in hand investigating the mudflats of Strangford Lough close to home, on a sunny day and surrounded by seabirds, Hans was connecting with nature. Elements of Sloane as a man were captured in this simple image: a fascination with the natural world around him, a commitment to observe, examine and focus, and collect specimens. It could have been any child lost in thought and wonder trying to make sense of the world around them, and perhaps this was the purpose of the poster. Bearing a message for Killyleagh's younger generation, life's journey, with keen observation and learning, can lead to greatness. It was also a reminder, to me, that it did not link to the man portrayed in formal surroundings with powdered wig, velvet jacket, lace shirt and stately, almost emotionless, demeanour. How might we connect the two images of the same person, bring them to life and tell the story of the intervening years?

On that same trip to Killyleagh my host had kindly arranged a behind-the-scenes visit to Killyleagh Castle, an opportunity offered by the Hamilton family whose lineage has been resident there for centuries. What was most memorable for me was the chance to sit in the library, with a mug of hot chocolate, each wall shelved with books and with a window onto the world. Here, through his father's employment to James Hamilton, young Hans Sloane was able to read and study, and develop his skills for collecting before he left Ireland to study botany, chemistry, anatomy and medicine. Since his first professional appointment, just a few years later, as physician in Jamaica where his own collecting began in earnest, Sloane began to bring the natural world together for reference, comparison, curiosity and use.

Although Sloane was a master collector in his own right, what sets him apart in the history of natural history was his masterful skill and commitment to collecting through the efforts of others and ultimately through the purchase of established, significant collections. This required organisation, financing and connections, all of which he had through his prosperous medical practice, marriage, business mind and advanced social standing. It is the legacy of Sloane's cast of collectors and collections that established the British Museum and, though today some might see such rewards as 'spoils' or 'plunder', this would not have been the perspective of the day.

Through the lens of the Sloane Herbarium, held at the Natural History Museum in London, this book captures not only Hans Sloane between young man and world-renowned naturalist, but the lives of explorers, discoverers and associates from all walks of life with whom he corresponded, traded and employed. Their lives, exploits and passions are no less important or fascinating, as the authors reveal, and their collections remain important reference material. A rapidly changing natural world makes such resources even more valuable and scientifically important, especially with the advance of imaging, digital and molecular technologies. Within and between the pages of Sloane's Herbarium are many new discoveries awaiting those willing to explore.

Tim Littlewood, *Executive Director of Science, Natural History Museum*

INTRODUCTION

MARK CARINE AND ROBERT HUXLEY

With more than five million specimens, the herbarium of the Natural History Museum, London, is one of the largest collections of preserved plant specimens in the world. At its core lies its founding collection, the herbarium assembled by the Irish-born physician, naturalist and collector Sir Hans Sloane. The 'Sloane Herbarium' is no ordinary herbarium. By modern standards it is small, comprising an estimated 120,000 specimens or around 2% of the total number of botanical specimens at the Museum. It is certainly not the oldest herbarium in existence and nor is it even the oldest herbarium at the Museum, but for its time – the period spanning the late 1680s when Sloane started to collect until his

Below Hans Sloane's herbarium in its climate-controlled home in the Natural History Museum's Darwin Centre with a portrait of Sloane in the foreground.

death in 1753 – its size, together with its geographical scope and the breadth of species that it contains, is without parallel.

During the course of his long life (he lived to be 93), Sloane amassed not only a herbarium but also many other objects including other natural history specimens, coins, medals, books, prints and manuscripts, which were all housed at his home, first in Bloomsbury Square and then at Chelsea Manor in London. In his will Sloane bequeathed his entire collection to the nation for the sum of £20,000, on condition that a public museum was created to house it. A national lottery was held to raise the necessary funds for such a museum and the British Museum in Bloomsbury, London, duly opened to the public in 1759. More than a century later, in 1881, the Natural History Museum opened in South Kensington housing the British Museum's ever-growing natural history collections, which included those of Sloane. The Sloane Herbarium is the largest surviving part of Sloane's natural history collections and is now housed in a purpose-built and climate-controlled room in the heart of the Darwin Centre.

Below Sloane was trained as an apothecary and a physician and was particularly interested in the uses of plants, including their medical uses. This drawer from his collection of Vegetable Substances includes plant parts used in medicine.

Specimens in the Museum's herbarium are preserved in many different ways: some are pickled, some are mounted on slides, while others are stored dry in boxes. However, the vast majority of specimens are dried and pressed and attached to sheets of paper, and those individual sheets are then filed in cabinets according to the group of plants that they belong to. It is a system common to herbaria worldwide and it has been the standard way of organizing herbaria since the great and influential Swedish naturalist Carl Linnaeus advocated the approach in the mid-1700s. One of its great advantages is its flexibility, since specimens can be easily rearranged if ideas about plant classification change, as they have done and continue to do so as our knowledge about plant diversity and evolutionary relationships advance.

Given its age, it is perhaps no surprise that the Sloane Herbarium is unusual by modern standards in the way it is arranged and catalogued. The specimens are mounted onto sheets, but not filed in cabinets. Rather, they are bound as pages or 'folios' into 335 large, leather-bound volumes or *horti sicci* (literally 'dry gardens'). Most herbaria from a similar period or older are arranged in this way but none on this scale. Sloane obtained his collections from many different people and within each volume the arrangement of specimens is typically dictated by the system adopted by its initial compiler. In some volumes the specimens are organized by species, genus and family, in others they are arranged alphabetically or geographically. In some cases, we find different systems in use within the same volume or no obvious arrangement whatsoever. Finding specimens of a particular species in such a diverse collection would be an almost impossible task were it not for the fact that Sloane was not only a great collector but also a great cataloguer. He catalogued all of his collections meticulously, and for his herbarium he devised an ingenious and simple cataloguing system. It made use primarily of *Historia Plantarum,* written by his great friend, the pioneering naturalist John Ray. Published in three volumes between 1686 and 1704, it was a work that sought to classify and describe all plant species then known. In the margins of Sloane's copy of this work, next to the name of each species, Sloane and his assistants recorded the herbarium volume and folio where specimens of that species could be found. Species not included in Ray's book – of which there were many – were listed in the page headers and footers. This remains the catalogue used today to locate plants in the Sloane Herbarium.

The challenge with using Ray's *Historia Plantarum* as a catalogue is that the plant names often bear little relation to those

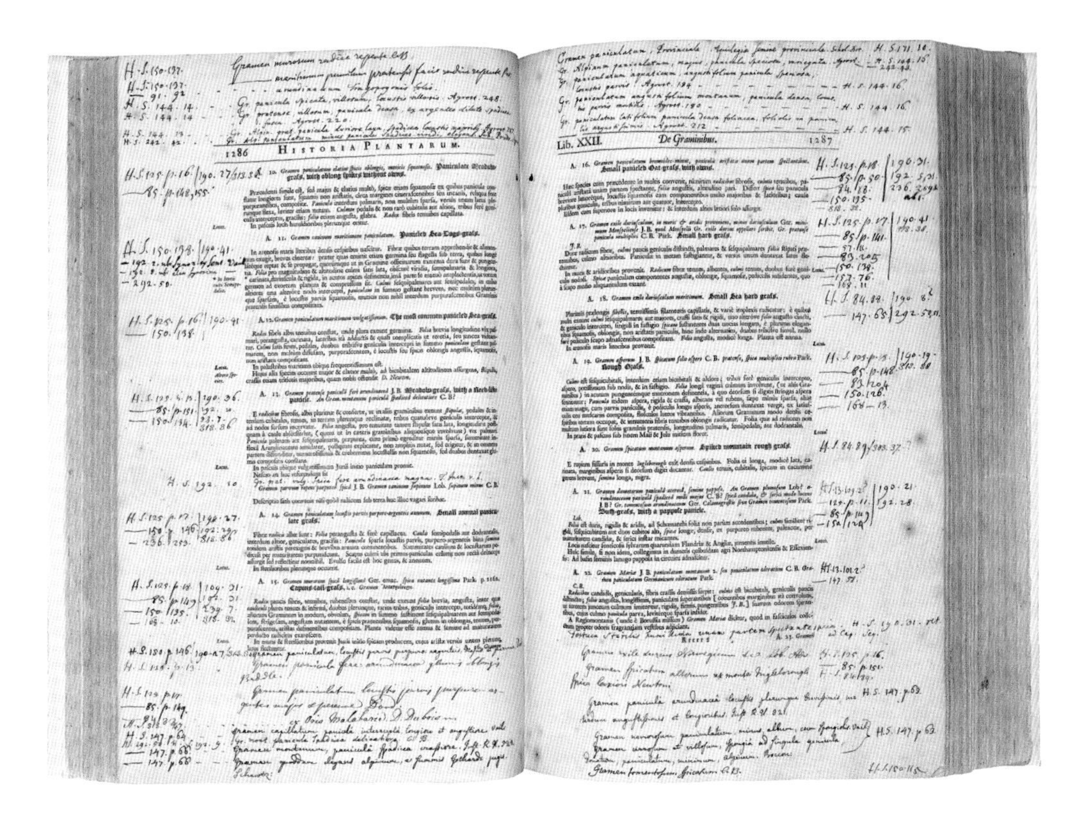

1286 HISTORIA PLANTARUM.

Lib. XXII. De Graminibus. 1287

currently used. Ray and his contemporaries used long, descriptive 'phrase names' in Latin to classify species. The name of the common poppy in Ray's *Historia Plantarum*, for example, was *Papaver laciniato folio, capitulo breviore glabro annuum Rhoeas dictum*. Today it is simply *Papaver rhoeas*. In 1753, the year Sloane died, Carl Linnaeus published his *Species Plantarum*, a new catalogue of plant species. In this he also used long, descriptive Latin phrase names but, in addition, provided a one-word 'trivial name' or 'specific epithet' for each in the margin. This was an indexer's paper-saving device that had been used before but following the publication of *Species Plantarum,* it became apparent that a combination of the genus name (*Papaver* in this example) and the indexing name (*rhoeas*) provided a much more efficient way to name plants than the

Above Sloane's copy of John Ray's *Historia Plantarum* was the catalogue to his herbarium. The annotations indicate the volume and folio in the herbarium on which each species can be found.

long phrase names used previously. And so, the modern binomial (two name) system of naming plants was born. The scientific name of the poppy became *Papavaer rhoeas*, the name still used to refer to it today.

Even though the Sloane Herbarium is distinctly 'pre-Linnaean', in both its arrangement in bound volumes and in the way the specimens are named and catalogued, its value has always been recognized. Linnaeus himself published descriptions and binomial names for species using illustrations that were drawn directly from specimens preserved in the Sloane Herbarium. Other botanists, including those trained by Linneaus, worked directly on the collections. The specimens they used in the Sloane Herbarium to describe new species remain scientifically important today as type specimens, an essential and unambiguous way to ensure that scientific names are correctly applied. But the significance of the Sloane Herbarium extends beyond these type specimens. A specimen in the herbarium is evidence that a particular species was growing at a particular place at a particular point in time. There are few herbaria that span the period covered by the Sloane Herbarium and it is consequently a unique source of data to inform our understanding of environmental change, evidenced by changes in the distribution (or even extinction) of plants over time. It also provides insights into how plants were being moved around the world and cultivated in places far removed from their native range and how European knowledge of plant diversity was developing during a period of major expansion in global networks of trade and empire. Today it is a collection studied by both biologists and historians alike.

Whilst Sloane was clearly the central figure in the assembly of his herbarium he was not working in isolation. Thanks to the efforts of James Britten and later James Edgar Dandy, the collectors and the collecting localities recorded in each volume of the Sloane Herbarium have been carefully catalogued. In 1958 Dandy published *The Sloane herbarium: an annotated list of the Horti sicci composing it; with biographical details of the principal contributors*, a catalogue that reveals more than 300 named contributors to the herbarium.

The nature of the contributions made by those individuals differed markedly. Some contributed just one or a handful of specimens. For example, Mrs Mary Lisle, one of only four women represented in the herbarium, contributed half a dozen plants collected in her garden in Crux Easton, Hampshire. They are nevertheless among the most beautifully preserved, with their

Left A specimen of larkspur collected by Mrs Mary Lisle of Crux Eaton, Hampshire. She corresponded with Sloane on how 'To preserve the Couler of Plants'.

colour still evident today. Other contributors were responsible for major parts of the herbarium. Sloane was not alone in amassing a large collection and using it to catalogue plant diversity and a number of his contemporaries were engaged in similar activities. Sloane exchanged specimens with them and sometimes even allowed them access to his herbarium. Furthermore, during his long life he purchased or was bequeathed the collections of several other fellow collectors and cataloguers. In 1696 he acquired the seven-volume herbarium of Christopher Merrett, whose *Pinax Rerum Naturalium Britannicarum* (1666) was one of the first lists of the flora, fauna and minerals of England. Other collections would follow. Sloane's largest acquisition was the herbarium of his friend, the apothecary James Petiver. Purchased in 1718, Petiver's herbarium today accounts for almost one third of the volumes in the Sloane Herbarium.

Gardeners were also important contributors. Many specimens cultivated in English, Irish and continental European gardens, often from seeds sent back from far-flung corners of the globe, are preserved in the herbarium. At a time of great innovation in horticulture, gardens such as those of Mary Somerset, Duchess of Beaufort, were a source of exotic species that were entirely new. Both the owners of the great gardens where plants were cultivated and the gardeners themselves became notable contributors.

Collectors of plants in the field were even more numerous as contributors. Not all field collectors sent specimens directly to Sloane; in some cases, specimens were incorporated only when Sloane acquired the herbaria of others such as that of Petiver. Ultimately though, those specimens contributed by collectors in the field enabled Sloane to assemble a herbarium with a remarkable global coverage encompassing more than 70 different countries and territories worldwide. Among the field collectors were merchants trading in a wide variety of commodities and, indeed, in enslaved people; there were dedicated plant collectors and members of the clergy, with William Sancroft, the 79th Archbishop of Canterbury, among them. But medically trained men – apothecaries and surgeons – were by far

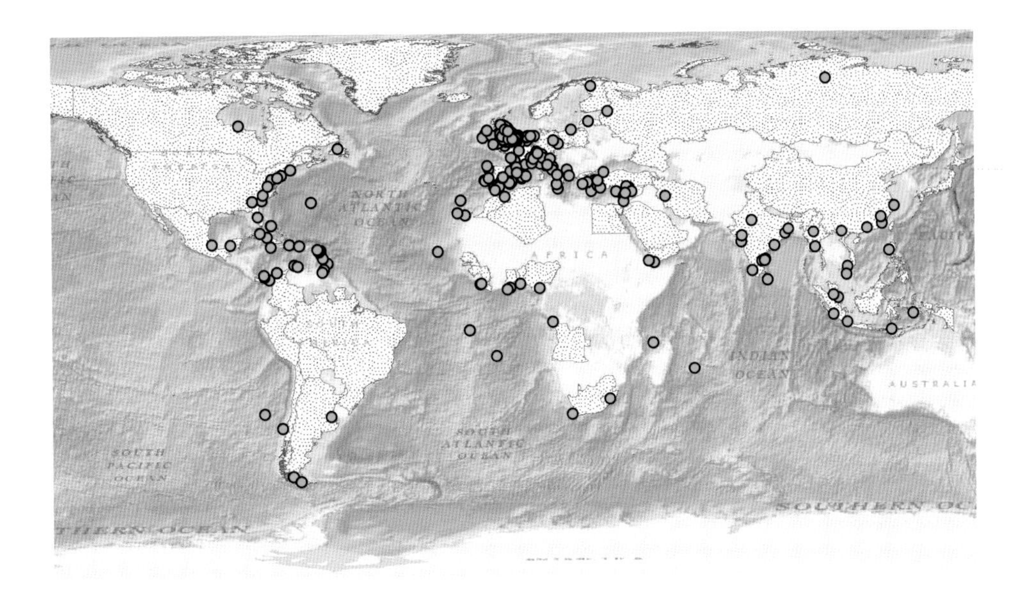

Above A map showing the countries (shaded in red) and specific localities (green dots) from where specimens in the Sloane Herbarium were collected.

the most numerous. Sloane himself trained as both an apothecary and as a physician and, for men like him, knowledge of plants was essential since it was plants that provided the raw materials for the remedies used in medical practice. Some surgeons and apothecaries travelled widely. Sloane himself travelled to Jamaica as a physician in the 1680s where he collected plants. Other medical practitioners were the contributors of specimens from some of the, then, most remote parts of the world including Japan, the Philippines and Tierra del Fuego.

Whilst the collections can be considered global in scope, specimens in the Sloane Herbarium originate from three main regions. Firstly, Britain and Europe, a region that is understandably the most comprehensively sampled with more than half of contributors to the herbarium active there. The second region, referred to in this book as 'the Atlantic', comprises the west coast of Africa, the central Atlantic Islands and the Caribbean and east coast of the Americas. Almost 30% of contributors to the herbarium were active in this area. The third region comprises Asia and the Pacific. Collectors in Asia were active from Turkey in the west to Japan in the east. Also included in this region are collections made along the Pacific coast of South America and in the Juan Fernandez Islands of the eastern Pacific. Collectors in Asia and the Pacific were less numerous, accounting for around 10% of contributors overall but they were nevertheless responsible for significantly expanding the geographical scope of the herbarium.

Who were those individuals collecting plants? Who were the gardeners and the great collectors and cataloguers? Why were they collecting plants? How did they contribute to this extraordinary collection and why are their contributions still significant today? Those are the questions that we explore in this book. We can consider only a selection of the contributors and the way in which we have categorized them is somewhat arbitrary – some contributors would fit into several of the themes. Nevertheless, it is a sample of contributors that, we hope, provides an insight into how Sir Hans Sloane's extraordinary herbarium was assembled and why, more than 250 years since Sloane's death, it is still important today.

Below A specimen of *Renealmia jamaicensis* var. *jamaicensis*, a member of the ginger family, that Sloane collected in Jamaica. It is mounted with a faithful drawing of the specimen by Edward Kikius who illustrated many of the specimens that Sloane collected during his Jamaican voyage. They were subsequently reproduced as life-sized plates in Sloane's *Natural History of Jamaica*.

Opposite A potato specimen (*Solanum tuberosum*) cultivated in Europe the 1660s. DNA extracted from this specimen has been analysed to understand how potatoes were domesticated in Europe since the 1600s.

Solanum tuberosum esculentum C.B.
papas peruvianorum
B. 4. 675.

THE GREAT COLLECTORS AND CATALOGUERS

MARK CARINE AND ROBERT HUXLEY

Sir Hans Sloane was a great collector and cataloguer of the natural world. His herbarium alone, unrivalled in size and scope, is testament to his ability and his commitment as a collector and is only a fraction of his entire collection. His skills in documenting the diversity of plants are evident from both the way in which he and his assistants carefully documented his herbarium using John Ray's *Historia Plantarum* and also in his published works, most notably his detailed and richly illustrated *A voyage to the islands Madera, Barbados, Nieves, S. Christophers and Jamaica*. Sloane's herbarium was an essential tool for both the descriptions and the illustrations of plants that he included in that work. Indeed, they could not have been produced without the specimens contained in his herbarium.

In a period when public museums were in their infancy, Sloane was not alone in assembling a large personal collection. For some 200 years individuals had been assembling collections in 'cabinets of curiosities', that were broad in scope and included objects from both the natural and man-made worlds. Designed to inform and to entertain, the exotic and the unusual were often highly prized, and the expansion of European activity in the Americas, Africa and Asia provided Sloane and his fellow collectors with a rich source of novel objects for their collections.

Hand-in-hand with the assembly of herbaria, scholars across Europe were seeking to make sense of the diversity of the natural world that those collections revealed. Through the careful examination of the characteristics of the specimens they contained, classifications were developed grouping specimens

Opposite A specimen on the St Helena Redwood (*Trochetiopsis erythroxylon*) from the herbarium of Leonard Plukenet.

into species, species into genera and genera into orders (or families). The approaches adopted by those cataloguers to classifying plants varied but collections were fundamental to this endeavour.

Many of Sloane's contemporaries as collectors and cataloguers – both in Britain and Europe – are represented in Sloane's herbarium today. In some cases, Sloane acquired entire collections, significantly enriching his own herbarium. In other cases, the contributions were more modest and involved the exchange of particular specimens. Other individuals contributed few specimens but through their efforts established classifications that were used by collectors, including Sloane, to bring order to their collections. Here, we account for five collectors and cataloguers who contributed to Sloane's herbarium.

Firstly, there is William Courten, a collector who assembled a remarkably diverse collection of objects that he housed in his own museum in rooms in Middle Temple, London. A complex character, known as both Courten and 'Charleton', who spent time in prison and often wrote his specimen labels in code, he created a herbarium that was greatly valued by Sloane who worked on it extensively after acquiring it. Courten was, without doubt, a significant early contributor to Sloane's collection.

Above Sloane acquired a herbarium comprising five volumes of plants collected by Herman Boerhaave, Professor of Botany and Medicine at Leiden University. The specimens, including these crocuses have decorative labels, typical of Dutch collections of the time.

James Petiver, another contributor, was once described as '...a person wretched in looks and actions, speaking very poor and deficient Latin and scarce able to string a few words together.' It is perhaps for this reason that his remarkable contribution to Sloane's herbarium – both in terms of the number of specimens and their geographical scope – has often been overlooked. A London apothecary by trade and, in contrast to Sloane, of limited financial means, it was his ability to cultivate a wide network of correspondents travelling across the globe that allowed him to assemble a herbarium that, at his death, rivalled that of his great friend Hans Sloane.

Petiver published on his plant collections, but his publications and his efforts to catalogue the natural world were more modest in scope and ambition than the remaining contributors. Leonard Plukenet – who in later life was appointed Royal Professor of Botany at Hampton Court – published a series of acclaimed works that described several thousands of species, many of them new. Sloane eventually acquired his herbarium, which was an important addition since the specimens contained in its 27 volumes formed the basis of Plukenet's published species descriptions. William Sherard's personal herbarium of 21,000 specimens never came into Sloane's possession since it was bequeathed to Oxford University where it is still housed today. Sherard was nevertheless a contributor to the Sloane Herbarium through the exchange of specimens between the two men. He embarked on an ambitious project to catalogue all plant names then described, a project that remained unfinished at his death. Whilst initially on cordial terms with Sloane, the stories of both Sherard and Plukenet reveal the conflicts and tensions that existed between at least some of those individuals engaged in assembling and classifying the great herbaria of the day.

The last two contributors, the Italian Pier Antonio Micheli and the Englishman John Ray, contributed few specimens to the herbarium. Indeed, of all the contributors we consider, Ray perhaps contributed fewer specimens than anyone else. Nevertheless, both Micheli and Ray were responsible for influential systems of classifying the diversity of the natural world that was being revealed by the great collections of the day. As we have seen already in the introduction, the system developed by Ray, a great friend of Sloane, was particularly important for the classification of the Sloane Herbarium.

Below Sloane's collection included many curiosities. This object, from his Vegetable Substances collection is described as *'The vertebrae of an ox through the hollow of wch passed an oak twigg wch grew above & below it swelling out'*. Sloane paid ten shillings for it – around £50 today.

WILLIAM COURTEN

SACHIKO KUSUKAWA

William Courten is one of the more enigmatic contributors to the Sloane Herbarium. He was the owner of a museum in Middle Temple in London, a fashionable visiting spot for ladies of the court, and an important source of information for many naturalists. Contemporaries praised it as a fine collection that included numerous shells, insects, medals, minerals, precious stones, objects in amber and dried specimens of fish and plants, as well as drawings of birds and plants. Courten was elected to a fellowship of the Royal Society, but did not take up the honour; he never published anything about botany or his collection, and was usually called Mr 'Charleton', rather than Courten. This is all the more surprising, given that his grandfather was the wealthy merchant Sir William Courten, who once operated a fleet of more than 20 ships with nearly 5,000 mariners. However, the family's fortunes had declined dramatically since then. A year after our William Courten was born, his father became insolvent and fled to Florence, leaving his family in England with losses estimated at the time to be about £500,000. The young William seems not to have been thrust into abject poverty, however: he grew up in relative comfort with his relatives in Northamptonshire, where he began collecting insects and butterflies. He even managed a shopping trip to London to purchase exotic and rare objects, amongst which were a Virginia woodpecker from Hester Tradescant, the widow of John Tradescant the Younger.

In 1670, however, he had to leave for Europe in order to avoid creditors of his family debt who were beginning to pursue him through the courts. In 1675 he met the philosopher John Locke in Lyons, and the two travelled to Montpellier. By February 1678, Courten had collected nearly 1,500 specimens of plants, about half of which were recorded as gathered in the mountains in

Opposite A folio from Courten's herbarium. It includes a number of botanical specimens – many of them fragmentary and with the labels partially written in cipher. It also includes the 'Indian writing on a reed' which he purchased in 1688.

49
Siliqua Sylvestris
Indian Writing upon
Wood & Reeds

Provence, L'Espérou, Bois de Gramont, La Colombiere, Castelneuf, Boutoniere, Capouladou and Lattes. Some of these may have been collected with Locke as the two men made their way to Montpellier, and it is possible that it was Locke who showed Courten how to make a herbarium. At some point Courten also copied out a method of drying out plants and a recipe for a 'German glue' for a herbarium from Adriaan Spieghel's *Opera Omnia* (1645). The remainder of Courten's plants listed in 1678 came from Montpellier – from the famous botanical garden of the university (Jardin du Roi) and the gardens of the physicians Louis Paul and Pierre Magnol. Courten drew up a list of these plants grouped by the locations at which they were collected, and identified their names using the standard reference books of the time, such as Caspar Bauhin's *Pinax Theatri Botanici*. Locke left Montpellier in February 1677, but Courten stayed on. He kept on collecting insects, shells, eggs, birds, lizards, corals, starfishes and a few more plants. He commissioned a local artist Guillaume Toulouze to paint watercolours of some items in his collection before leaving Montpellier for Paris in 1682, where he was briefly imprisoned on suspicion of cavorting with Protestants at Montpellier. On release, he stayed on in Paris, collecting medals and prints, but not, it seems, natural objects. Instead, he commissioned watercolour drawings of plants and birds on vellum from the famous painter Nicolas Robert. Courten probably met Sir Hans Sloane in Paris in 1683, when Sloane was studying at the Jardin Royal des Plantes and the Hôpital de la Charité. Soon afterwards, feeling that the threat from his creditors was abating, Courten decided to return to England. He would assume the name of 'Charleton', to evade other possible creditors.

Courten rented rooms at Middle Temple in London, where the antiquary Elias Ashmole, the founder of the Ashmolean Museum in Oxford, had kept his collection until a fire destroyed it in 1679. By 1685, Courten was showing his collection to visitors. There was now plenty of opportunity to augment his collection. Lists of purchases made through his perambulation of London indicate that at pubs, the docks and street corners, from sailors, hawkers, widows

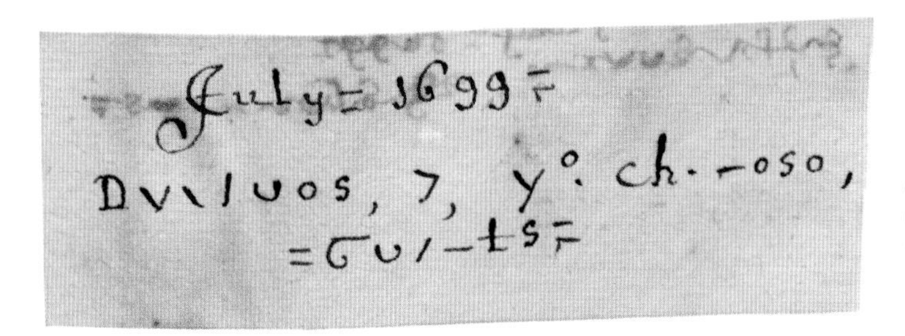

Left and below Courten's cipher: 'Doubles of the Chinese plants' (left) and 'Plants from China given me by Mr Maidstone, 72 plants, a sort of lilly (sic)' (below).

and goldsmiths, Courten purchased rare and exotic things from around the globe. He obtained a root of 'gensum' from Persia from a ship that had come in from East India, and one Mrs Bonfield sold him the feet, a wing and a tail of 'a sort of owl from Russia'. Bits and pieces of mineral ores as well as parts of fauna (beaks, legs, shells) and flora (nuts and seeds) that could endure lengthy voyages across the seas could readily be acquired around London. In his lists, Courten recorded the prices he paid for objects as well as the prices at which he was willing to part with them. Those objects he was willing to sell were marked with 'D', for 'Double'. Given his financial constraints, Courten was resourceful in augmenting his collection through these doubles – often by exchanging an object for something else. He wrote his accounts in cipher, perhaps to protect his sources, or the identity of the objects.

On returning to London, Courten appears to have gone on a few herb-gathering trips and plants collected by him from Hampsted Heath, Sussex, Dover and Carshalton (Surrey) have survived. He was, however, more active in sourcing exotic plants. Instead of relying on what other Londoners could get hold of, Courten decided to ask those about to go overseas to bring things back for him. He supplied them with tools (scissors, pins, hooks, 'spirits of wine') and instructions for what to collect, as was the case with James Reed. Reed's destination, Barbadoes, must have rung poignant for Courten, whose grandfather had funded the first English settlement there. Coutern had hoped to recover rights to the island but had been unsuccessful. Travelling surgeons James Cuninghame and George Handisyd brought him plants from China and the Straights of Magellan, respectively. Naturalist scholars also sent Courten exotic gifts: William Sherard gave him parcels of seeds from Africa and the East Indies, Leonard Plukenet plant specimens from east and west India, Samuel Doody seeds from Fort St George, Jacob Bobart the Younger a '*cucumis Africanus Echinatus*' and James Petiver 'a twisted gourd'. In return, Courten offered exotic seeds and plant specimens. Several of these men, including Sloane, belonged to a club that met at the Temple Coffee House or at Courten's rooms nearby to discuss botanical matters. The names sporadically mentioned in Courten's herbaria indicate the network of people with whom Courten transacted botanical material.

Courten's herbaria contained not just plants. Glued on to a sheet, next to some plant specimens was a palm leaf with Balinese script, which is likely the 'Indian writing on a reed' purchased in 1688, and a head of a woodpecker from China.

Courten's specimens are carefully and neatly glued on to paper, though their names are not always established. Some sheets of paper contain just one kind of plant, such as grasses, or plants from one location. Since several of these are marked as doubles or

Left The head of a woodpecker (probably a *Dinopium* sp.) mounted among pressed plants in Courten's herbarium. The head, like its associated plant specimens, was probably collected in Southeast Asia about 1698–1699 by Nathanael Maidstone, an East India Company trader.

'for sale', the grouping by kind or by origin should be seen as categories by which other collectors commonly sought specimens. There was also some reorganization across his herbaria, as Courten cut out plants from their original sheets or unglued them and placed them onto a new sheet, creating a new collage. This could have been an opportunity to bring together past and present plants for study, but the organizing principles here are unclear. There is no evidence that Courten thought deeply about taxonomy. He sometimes mixed his own plants with those given to him by Sloane, perhaps indicating a sense of affinity with him. The care and neatness with which the plants were rearranged suggest some sense of aesthetic, revelling in the variety of forms nature could produce, a trait that can be appreciated from the arrangement of his earliest specimens.

Above Grasses from France and Spain, marked by Courten, 'The choicest of my gramens, for sale'.

Several contemporary naturalists found something worth their attention in Courten's collection. Among others who expressed their gratitude to 'Mr Charlteon' were Petiver, Plukenet, the Oxford professor Robert Morison and John Ray. Courten never married, and left his cherished collection, which he valued at £7,000–8,000, to Sloane. It was the first entire collection that had come into Sloane's possession. Sloane meticulously went through Courten's herbaria, cross-referencing the plant names with Ray's *Historia Plantarum*, which he used as an index to his herbarium. Courten's passion for prints, coins and drawings of natural objects may well have rubbed off on Sloane. Courten's was the collection that Sloane cut his teeth on, and it launched his career as a collector.

JAMES PETIVER

RICHARD COULTON AND CHARLES E. JARVIS

Of all the collectors represented in Sir Hans Sloane's extraordinary herbarium, no one contributed to its unprecedented size and range more significantly than his great friend James Petiver. A London apothecary from a modest background, Petiver became prominent within Sloane's circle shortly after the physician's return from Jamaica in 1689. United by their love of botany, Sloane and Petiver regularly exchanged intelligence, contacts and specimens. In the 1690s they co-ordinated sociable meetings of London plant-lovers, notably at the Temple Coffee House, and advanced the cause of natural history within the Royal Society (where Sloane helped to secure Petiver's election as Fellow in 1695). Despite Sloane's social advancement as the years elapsed, the two men remained on good terms until Petiver's death in 1718. Sloane publicly officiated as a ceremonial pall-bearer at his funeral.

When Petiver died, not only did Sloane lose a key companion and collaborator, he gained an opportunity to acquire the most diverse and probably the most extensive botanical collection in Great Britain. Petiver's plants now constitute 106 of the 334 volumes – almost a third – of the Sloane Herbarium. They evidently formed an even higher proportion when Sloane purchased them (along with the rest of Petiver's museum and manuscripts), for around this time he claimed to own '200 large volumes of dried Samples of Plants' including those from Petiver's estate. In the introduction to the second volume of his *Natural History of Jamaica* (1725), Sloane posthumously celebrated Petiver as a man of great 'Understanding in Natural History' who had amassed a 'greater Quantity' of specimens 'than any Man before him'. Yet he also complained that his friend did not take appropriate 'Care' of his collections: being 'injured by Dust, Insects, Rain, &c.' they had necessitated painstaking and urgent conservation on Sloane's part.

Opposite An English moss in Petiver's herbarium, 'Mr Doody's hairy Goldilocks' (*Grimmia pulvinata*) is mounted with text cut from one of Petiver's own publications, with his added handwritten references to earlier accounts of the species.

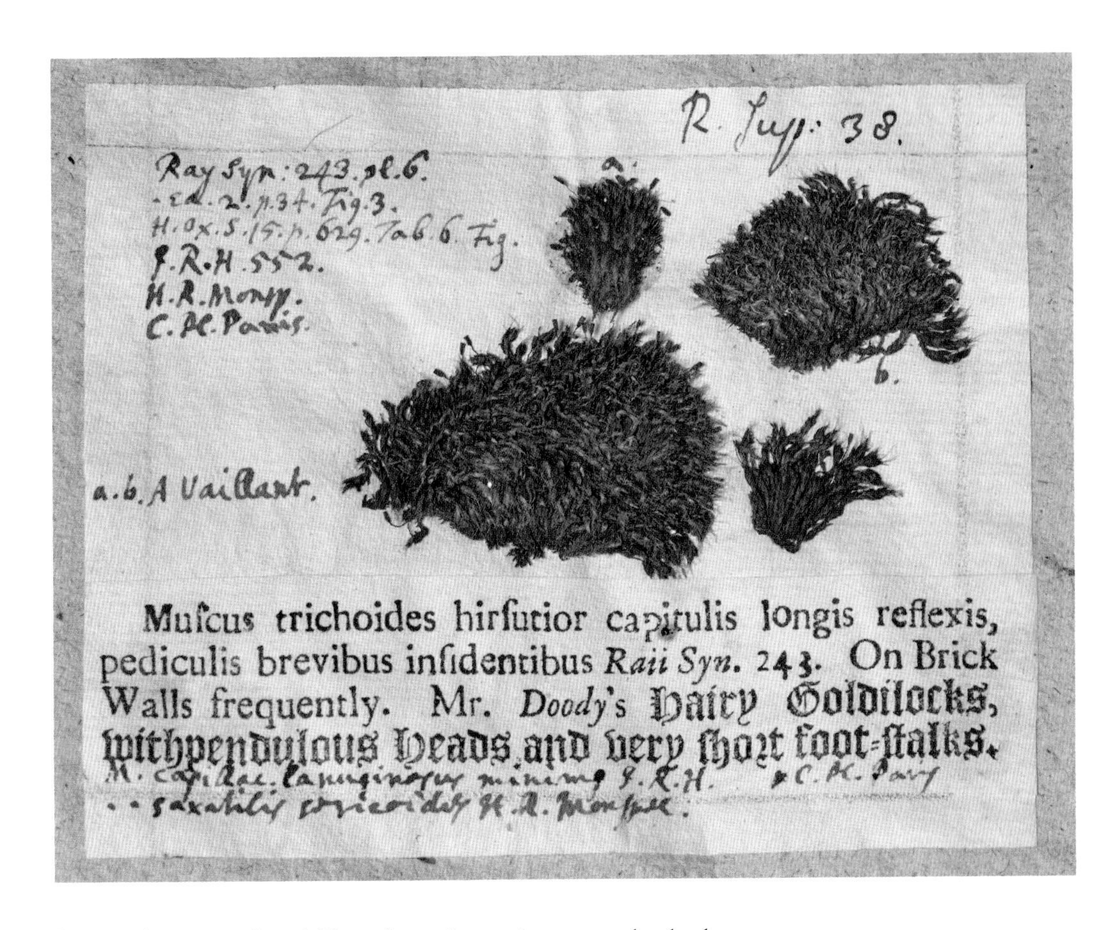

History has remembered Sloane's condemnation more clearly than his praise, and as a consequence Petiver has tended to be dismissed as scatty, disorganized and unlearned. Nonetheless, recent botanical and entomological research shows that despite his limited means and education, Petiver was an excellent and original naturalist, with considerable taxonomic abilities.

So how did James Petiver – an urban tradesman of the middling sort – amass a global plant collection that was so coveted by Sloane, and in which were represented new species from the five known continents of the world? The answers lie in Petiver's fastidiousness, focus and friendships. Where Sloane leveraged his personal wealth to purchase wholesale collections of rarities, and cultivated his social prestige to procure assistance from individuals of learning and affluence, Petiver's remarkable achievement was to engage a wholly new class of natural historians in his service: ordinary people. A ruthless and relentless networker, Petiver built connections with all kinds of observant travellers, domestic and international, whom he persuaded to gather specimens on his behalf. Addressing men (and occasionally women) of his own social status, Petiver created detailed written instructions – at first in manuscript, and later printed at his own expense – that provided his contacts with requisite technical directions for the fieldwork of obtaining, documenting, preserving and transmitting samples of plants, insects and shells. Curiosities from all over the planet thereby flooded into his apothecary's shop at the sign of the White Cross, on Aldersgate Street in the City of London.

All this activity culminated in a herbarium comprising an estimated 30,000–35,000 distinct gatherings (as opposed to individual specimens), contributed by more than 160 named collectors or suppliers. Many who feature elsewhere in this volume owe their inclusion to Petiver's assiduousness in enlisting them for his project. Among his contributors were English clerics such as Adam Buddle, fellow apothecaries like Samuel Dale, physicians such as the Yorkshire-based Richard Richardson and the Scot Patrick Blair, and the natural philosopher and Keeper of Oxford's Ashmolean Museum, Welshman Edward Lhwyd. Moreover, Petiver was himself a keen field biologist who made regular forays in search of interesting plants, chiefly around London and the southeast of England, often in the company of friends such as Samuel Doody, Keeper of the Society of Apothecaries' garden at Chelsea (where Petiver was later appointed Demonstrator). Petiver also studied plants under cultivation in various London gardens, including those of the Bishop of London at Fulham Palace, and Mary Somerset, Duchess of Beaufort, in Chelsea.

However, it is the geographical range from which Petiver acquired his specimens that is particularly staggering for the period. As might be expected, much botanical material arrived through Petiver's European contacts such as Sebastian Vaillant, Regius Professor of Botany in Paris; the Danzig (Gdansk) merchant Jacob Breyne; the Barcelona apothecaries Joan Salvador i Riera and Jaume Salvador i Pedrol; and the Tuscan abbot Bruno Tozzi. The arrival of material from the New World was facilitated not least by the slave trade, with ship's surgeons such as Richard Planer supplying specimens from ports on either side of the Atlantic, in both West Africa and Cartagena (Colombia). From English colonies in North America, Petiver obtained material via clergymen such as Hugh Jones in Maryland and John Banister in Virginia, and plants also arrived from the short-lived Scottish settlement at Darien (Panama) via the surgeon Archibald Stewart. Much material came from the botanically rich Cape of Good Hope, directly from collectors such as Henry Oldenland, a Dane who was in charge of the Government Garden, and the Dutch East India Company employee John Starrenburgh. Specimens also came indirectly from the Netherlands via Professor Frederik Ruysch from Amsterdam. The trade routes to Asia produced pharmaceutically interesting material from India via the East India Company surgeons Samuel Browne and his successor Edward Bulkley; very significant collections came from Indonesia, Vietnam and China, primarily through the enthusiasm of the Scottish surgeon James Cuninghame; along with

Below A specimen of a clubmoss (*Selaginella exaltata*) collected at the ill-fated Scottish settlement at Darien by the surgeon Archbald Stewart, beneath which is mounted a cut-out copy of Petiver's published description.

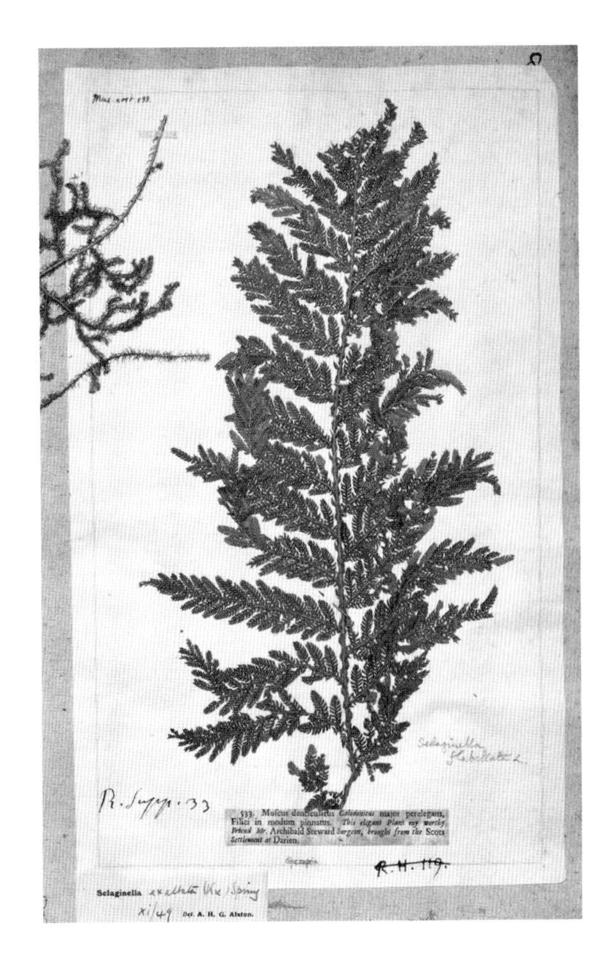

specimens and drawings supplied by the Moravian lay brother Georg Kamel from the Philippines.

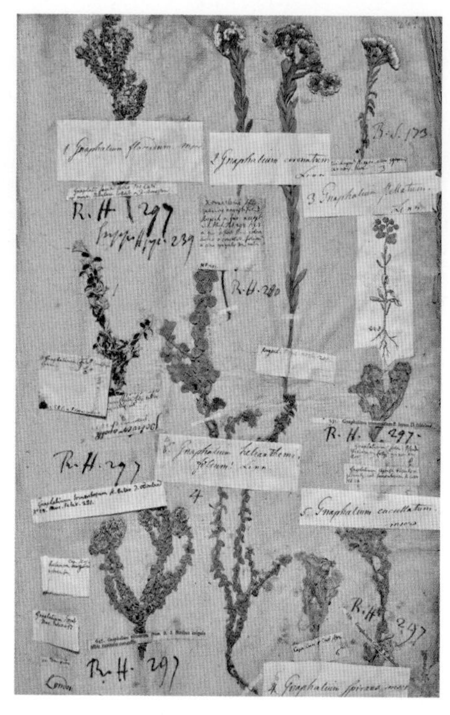

Left A page from Petiver's herbarium showing similar species from the Cape of Good Hope mounted together. The labels indicate that they came from four different collectors (James Cuninghame, Henry Oldenland, Frederik Ruysch and George Stonestreet).

At first sight the volumes in Petiver's herbarium can appear rather less impressive than those prepared by Sloane himself and by other collectors such as Leonard Plukenet. Petiver's volumes are generally smaller and more variable in size than those of his wealthier contemporaries. They typically contain pages of coarse, unbleached, lightweight paper, each sheet of which can bear upwards of 12 gatherings. However, the organization of the specimens is importantly distinctive. Where possible, Petiver began with a geographical layout within which specimens were arranged following John Ray's classificatory system. This groups similar species together (in contrast to Sloane's collector-based and Plukenet's alphabetical arrangements), making Petiver's collections far more taxonomically informative. Petiver was also conscientious in retaining collectors' original labels alongside his own (which often indicate a given specimen's source). A single page, therefore, can carry specimens of the same or similar species, collected by different people at different times, and enable valuable side-by-side comparison. As with material indexed across the herbarium, Petiver's specimens and the corresponding species descriptions in Sloane's copy of Ray's *Historia Plantarum* also benefit from the reciprocal annotations of Sloane and his amanuenses.

The thousands of specimens that Petiver painstakingly acquired did not languish unseen in his lifetime. Alongside the singular innovation of selling sets of dried specimens with printed labels, Petiver published extensively on his plant collections, most notably in his *Musei Petiveriani* (1695–1703), which included 800 short descriptions of plants, and the more extensively illustrated *Gazophylacii naturae et artis* (1700–1711). Additionally, nearly 30 articles appeared in the *Philosophical Transactions of the Royal Society* between 1695 and 1716. Petiver's publications usually indicate the provenance of the specific material they describe: the source specimens themselves – frequently annotated by Petiver in ways that make the link with the relevant publication explicit – can often still be located in the herbarium today. In *Species Plantarum* (1753), Carl Linnaeus cited many of Petiver's published descriptions and illustrations, some of which serve as nomenclatural types for their corresponding binomial names (e.g. the fern *Adiantum philippense*). Petiver is commemorated in the genus *Petiveria*.

Above A plate from one of Petiver's publications depicting a range of unrelated plants, butterflies and moths from various parts of the world, along with a single bird.

PIER ANTONIO MICHELI

ROBERT HUXLEY

...surely the young man who had found it if he continued with the same fervour would become a great botanist...

So said the great French botanist Joseph Pitton de Tournefort when confronted with a plant specimen collected in the high Alps by a young Florentine of humble birth. Tournefort felt that only someone driven by a great passion for botany would be daring enough to climb these high and terrifying mountains. His prediction was realized and the young man in question, Pier Antonio Micheli, was to become a highly able and respected botanist. He became better known for his work as a mycologist, his ground-breaking fungal discoveries bringing him the unofficial title of 'father of mycology'. Micheli's success in both fields was in spite of the prejudices that were associated with a person's educational and social background at the time. During his 57 years he amassed a large collection of fossils, invertebrates and a substantial herbarium now housed in the Museo di Storia Naturale di Firenze, the natural history museum in Florence. Through his active correspondence with James Petiver there are many examples of his collections distributed through the Sloane Herbarium and his beautifully illustrated book *Nova Plantarum Genera* was to inspire later botanists and mycologists including Carl Linnaeus.

Micheli was born in Florence in 1679 into a humble background. His parents were financially unable to support a formal education for their son and encouraged the young Pier Antonio to take up an apprenticeship as a bookbinder. However, the young man's passion was botany and he spent as much time as he could collecting and identifying the plants and fungi of his native countryside. It was while working amongst the rare and ancient books of his employer Ottavio Felice Bonaiuti that he came across *Commentaries on Dioscorides*, a botanical text published in 1585 by the botanist and physician Pietro Andrea Mathioli. It was a practical scientific treatise intended for daily use by physicians and herbalists and Micheli delighted in returning home with his day's collections to check his identifications against Mathioli's illustrations.

Although his father had initially discouraged his son's botanical studies and travels he seems to have been impressed by the expertise that his son had developed – and by the money he was able to earn by obtaining medicinal plants for local herbalists. Micheli's first steps out of obscurity and into the botanical limelight came through his friendships with three fathers from a monastery some 40 km (25 miles) from his home. Virgilio Falugi, Biagio Biagi and Bruno Tozzi were all accomplished botanists and with their help Micheli honed his identification skills and taught himself Latin, an essential skill for any scholar at the time but inaccessible to most people.

Through his friendship with the fathers, Micheli became known to the Florentine nobility including Giuseppe Del Papa the court physician to Cosimo III de' Medici, Grand Duke of Tuscany. This can be regarded as Micheli's 'big break' as Del Papa introduced him personally to the Grand Duke who took the unprecedented step of inviting this young commoner to lunch where they discussed matters botanical. Cosimo was so impressed by the Florentine's knowledge and skill that in 1706 he appointed him as his court botanist and assistant keeper of the Garden of Simples (plants of medicinal value) in Pisa. To invite to lunch and then give a high-ranking position to a lowly born 26-year-old is a testament to Micheli's abilities and the esteem in which he was held.

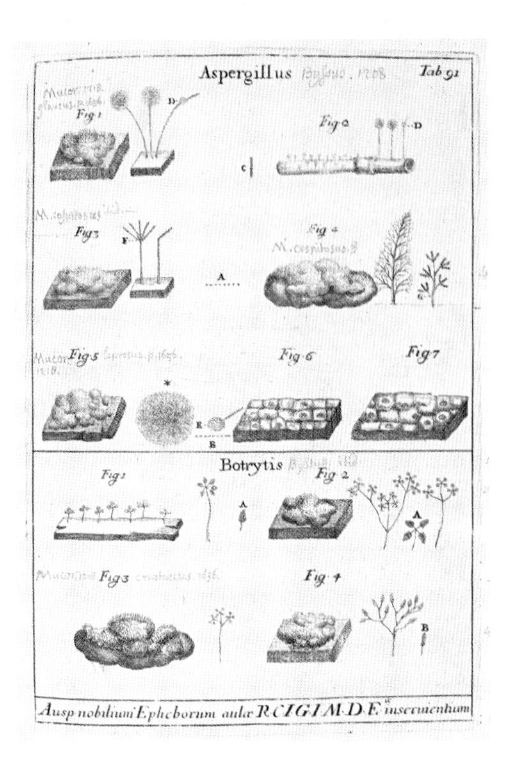

Above Plate from Micheli's *Nova Plantarum Genera* illustrating two fungal genera first named by him: *Botrytis* and *Aspergillus*. *Aspergillus* reminded Micheli of an aspergillum or holy water sprinkler.

Within a year Micheli had become Professor of Botany at the University of Pisa. Supplied with the latest botanical texts by Cosimo, effectively equivalent to a state-of-the-art laboratory today, he began to develop his own theories and compile his own great work on plants and fungi.

One of the texts supplied by Cosimo was Tournefort's *Institutiones rei herbariae* which laid out for the first time a robust definition of genera and outlined his views on species, a key topic for discussion at the time. This provided the basis for Micheli's own concepts. The classifications of Tournefort, Micheli and their later admirer Linnaeus were pragmatic and easy to use. Micheli employed his classification along with his excellent illustrations in the first and only published part of his *Nova Plantarum Genera* of 1729, acknowledged to be one of the best manuals of plant classification and identification of its time. This beautiful book was dedicated to its financier, Gian-Gastone de' Medici, son of Cosimo, and is packed with high-quality illustrations from 108 copperplates.

Left Dog-stinkhorn fungus illustrated in Micheli's *Nova Plantarum Genera*. This illustration is the type for Linnaeus's name for this species of fungus, *Phallus impudicus*.

Right Italian alpine plants (mainly *Gentian* and *Gentianella* species) collected by Pier Antonio Micheli and sent to James Petiver.

Nova Plantarum Genera included descriptions of some 1,900 plant and fungal species. The majority were so-called 'lower plants' or 'cryptogams' – a bucket term for a range of sometimes very different organisms – such as ferns, mosses and liverworts, and fungi including lichens. Micheli was the first to classify these groups of organisms and he described and named many of the fungal genera recognized today such as *Aspergillus*, species of which are used to ferment soy sauce, and *Botrytis*, responsible for

24
R.H. 717
R.H. 716
Gentiana alpina latifolia magno flore CB. 187.1.
R.H. 718
Vaillant.
R.H. 719
R.H. 718

diseases in grape and other fruit crops. Many of the fungal illustrations in this work later served as types for Linnaeus' binomial names, including well-known fungi such as *Phallus impudicus*, the dog stinkhorn.

Armed with two primitive microscopes, a keen eye and a scientific approach, Micheli created a window into the then unknown world of lichens and other fungi and among his pioneering discoveries showed that fungi reproduced by spores. He followed up his microscopic observations of their fruiting bodies with a series of elegant experiments. By transferring spores of a mould fungus with a fine brush to freshly cut slices of melon, he observed that the 'new' mould appearing on the melon bore exactly same fruiting bodies as the 'parent', thus challenging the common belief of the time that moulds arose by spontaneous generation from rotting matter.

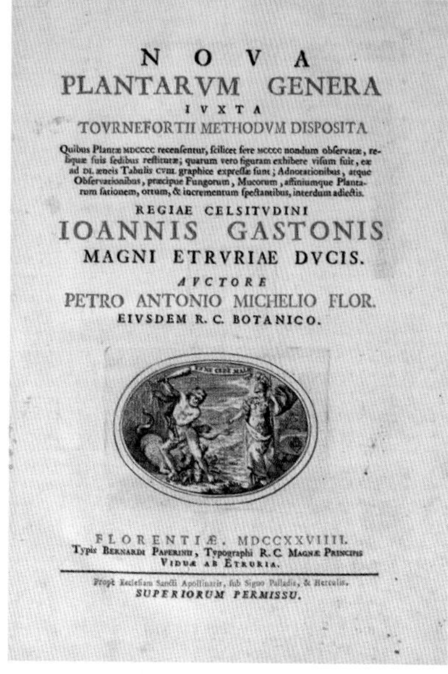
NOVA
PLANTARVM GENERA
IVXTA
TOVRNEFORTII METHODVM DISPOSITA
Quibus Plantæ MDCCCC recensentur, scilicet fere MCCCC nondum observatæ, reliquæ suis sedibus restitutæ; quarum vero figuram exhibere visum fuit, eæ ad DL æneis Tabulis CVIII. graphice expressæ sunt; Adnotationibus, atque Observationibus, præcipue Fungorum, Mucorum, affiniumque Plantarum sationem, ortum, & incrementum spectantibus, interdum adiectis.
REGIAE CELSITVDINI
IOANNIS GASTONIS
MAGNI ETRVRIAE DVCIS.
AVCTORE
PETRO ANTONIO MICHELIO FLOR.
EIVSDEM R. C. BOTANICO.
FLORENTIÆ. MDCCXXVIIII.
Typis Bernardi Paperinii, Typographi R. C. Magnæ Principis Viduæ ab Etruria.
Prope Ecclesiam Sancti Apollinaris, sub Signo Palladis, & Herculis.
SUPERIORUM PERMISSU.

Above Frontispiece of Pier Antonio Micheli's *Nova Plantarum Genera*. The dedication is to Gian-Gastone de' Medici, Grand Duke of Tuscany.

Micheli's observations were not confined to plants and fungi; he also had an interest in geology. On one plant-hunting trip to the Adriatic islands he noticed that the local rocks were similar to those on Mount Vesuvius and correctly concluded that the islands had once been volcanos. He excelled as a field botanist and collector. He was meticulous in his recording of observations, the logging of his journeys and the collection and preparation of specimens. On the labels of his early specimens Micheli recorded very specific information on the collection locality – the names of owners of market gardens and farms and even the bird-catching tunnel of his early supporters, the Florentine gentry.

It was his post of assistant to the director of the gardens at Pisa that provided him with the remit to roam wider, journeying extensively to find new plants for the gardens in Pisa and also Florence. Micheli usually travelled alone, covering most of Italy and then on to Tyrol, Bohemia, Prussia, Thuringia and Istria. As Tournefort had first observed, Micheli would stride alone across the most rugged terrain without any regard for safety or comfort to obtain his precious plants and fungi.

By the age of 30 Micheli was recognized across Europe for his abilities and collections but was not immune to criticism and some snobbery. Fellow alpine collector and naturalist Albrecht Haller called him illiterate, implying that he had no Latin. James Sherard, brother of the great botanist William, was a harsh critic as was German botanist Jacob Dillenius. However, William Sherard, was to be a key factor in promoting Micheli's career not only praising him as the greatest botanist of his day to the Grand Duke but also becoming his patron. Through the influence of William Sherard and others, Micheli built a network of contacts with which he corresponded and exchanged specimens. Like many of the botanists described in this book, he exchanged specimens with James Petiver and it is through him that Micheli's specimens can be found in the Sloane Herbarium.

The first part of *Nova Plantarum Genera* was the only part completed. Micheli had been collecting algae intending to add pages on them. Sadly after a final collecting trip to Monte Baldo (near Lake Garda), Micheli contracted pleurisy and died. He was buried in his doctoral robes in 1737.

Micheli's personal herbarium was bought by his student Giovanni Targioni Tozzetti and in 1845 sold by Giovani's grandson to the Musueo di Fiscia e Stria Naturale of Florence. His collections were integrated into the Micheli-Targioni herbarium and remain there to be used by botanists to this day along with the specimens distributed around European collections such as those found in the Sloane Herbarium.

Micheli's lowly background would never allow him to reach the heights of others but within his modest posts at Florence and Pisa he was able to make his mark as a highly skilled mycologist and botanist, and amass large collections that are valuable to this day. He is commemorated by a statue in the Uffizi Gallery and two street names in Florence and by the genus of trees and shrubs, *Michelia* (Magnoliaceae) named in his honour by Linnaeus.

Below Statue of Pier Antonio Micheli in the Uffizi Gallery, Florence.

WILLIAM SHERARD

STEPHEN HARRIS

William Sherard is poorly known compared to his contemporaries such as John Ray, Sir Hans Sloane and James Petiver. However, he was not merely a witness to developments in early-eighteenth-century European botany; Sherard was one of the architects. The Swedish botanist and 'father of plant taxonomy' Carl Linnaeus described him as 'Botanicus Magnus'; to the naturalist and Linnaean 'apostle' Fredrik Hasselqvist he was the 'Regent of the Botanic world'; and to the editor of Richard Richardson's letters he was 'the Sir Joseph Banks of his day'.

Sherard was born in Bushby, the eldest son of a Leicestershire landowner. In 1677, he was awarded a fellowship at St John's College, Oxford, where he read law and graduated in 1683. As a student Sherard began a friendship with the second keeper of the Oxford Physic Garden, Jacob Bobart the Younger, which continued for the rest of Bobart's life. Bobart was instrumental in developing Sherard's botanical interests which, in the latter part of the century, were augmented under the tutelage of Joseph Pitton de Tournefort at the Jardin du Roi in Paris and Paul Hermann and Herman Boerhaave at Leiden. Following Hermann's death Sherard helped Sloane purchase natural history specimens and books from the estate, and performed similar functions for him in the early eighteenth century.

On returning to England Sherard became travelling companion or tutor to wealthy individuals such as Wriothesley Russell, Second Duke of Bedford, the Irish landowner and renowned gardener Arthur Rawdon, the prominent Whig politician Charles Townshend and the grandson of Mary Somerset, Duchess of Beaufort, who also features in this book. In 1703 Sherard was appointed to the prestigious, and well-remunerated, position of consul at Smyrna (modern İzmir, Turkey) for the Levant Company.

Among his duties, and other interests, he continued his botanical pursuits, sending Smyrnaean specimens to Sloane. He returned to England in 1717, a wealthy man. A year later he was elected a Fellow of the Royal Society.

Sherard was based in London during the final years of his life. However, he continued to travel to Europe, cultivate botanical relationships and acquire, maintain and renew friendships. In particular, he employed the talented young Hessian botanist Johann Jacob Dillenius as his assistant. When Sherard died in 1728, he endowed the chair of botany at the University of Oxford and made a bequest of his herbarium and library. One of the three conditions attached to the endowment was that Dillenius was to be the chair's first incumbent. Sherard is buried in Eltham, the home village of his brother James Sherard.

In 1816 James Edward Smith, founder of the Linnean Society of London and purchaser of Linnaeus' herbarium, praised Sherard's herbarium as 'perhaps, except that of Linnaeus, the most ample, authentic, and valuable botanical record in the world' and, at the start of the eighteenth century, it was being praised as 'the most complete collection of North American plants'.

Above The specimen of the cocoa tree (*Theobroma cacao*), collected in Jamaica, from William Sherard's personal herbarium that was pressed on the drying sheet shown on p.43.

Sherard's herbarium in Oxford comprises approximately 21,000 sheets. Unusually for the period, it never appears to have been bound in book form, unlike the Sloane Herbarium, or the 80 elephant-folio volumes that made up the herbarium of Sherard's friend, Charles Dubois. Furthermore, Sherard appears not to have acquired entire herbaria of other collectors, in contrast to both Sloane and Dubois. Specimens in the Sherard Herbarium, although typically with limited information about collector and collection date and locality, were acquired from many different sources. Geographically, specimens come from throughout Europe, from southern South America to Greenland, South Africa to North Africa, Turkey through Central and South Asia to East Asia and even Australia.

Above Portrait identified as William Sherard, in oil by an unknown artist.

Sherard's herbarium appears to have been a tool for helping him to complete the *Pinax*, a revision of Caspar Bauhin's *Pinax theatri botanici* (1623). By the middle of the seventeenth century Bauhin's list was out of date and tens of thousands of polynomials were being used to communicate about plants. However, people did not know whether they were talking about the same species or different species. Sherard's *Pinax* would be a catalogue of the world's plant names that identified and linked all of the synonyms. It was a vast undertaking.

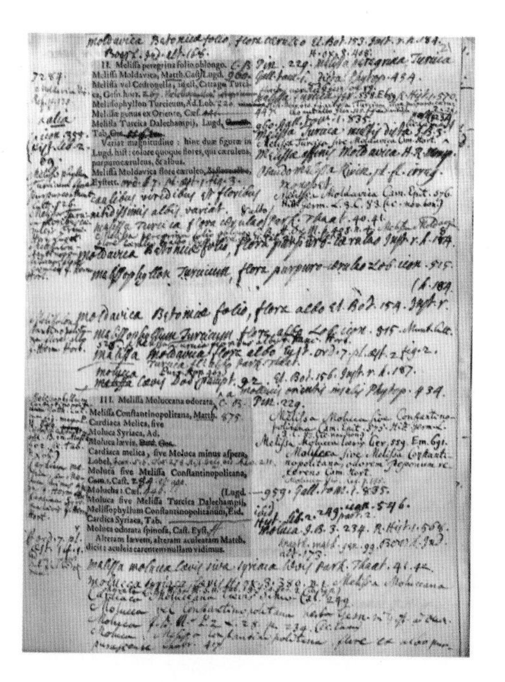

II. Meliſſa peregrina folio oblongo.
Meliſſa Moldavica, Matth. Caſt. Lugd.
Meliſſophyllon Turcicum, Ad. Lob.
Meliſſa Turcica Dalechampii, Lugd.
Variat magnitudine: hinc duæ figuræ in Lugd. hiſt: colore quoque flores, qui cæruleus, purpureocæruleus, & albus.
Meliſſa Moldavica flore cæruleo,
Eyſtet.

III. Meliſſa Moluccana odorata,
Meliſſa Conſtantinopolitana, Matth.
Cardiaca Melica, ſive
Moluca Syriaca, Ad.
Moluca lævis,
Cardiaca melica, ſive Moluca minus aſpera, Lobel.
Moluca ſive Meliſſa Conſtantinopolitana, Cam. Caſt.
Moluccha: Caſt. (Lugd.
Moluca ſive Meliſſa Turcica Dalechampii,
Meliſſophyllum Conſtantinopolitanum, Eid.
Cardica Syriaca, Tab.
Moluca odorata ſpinoſa, Caſt. Eyſt.
Alteram lævem, alteram aculeatam Matth. dicit: aculeis carentem nullam vidimus.

Left Typical page from Sherard's *Pinax*, with printed clippings from Bauhin's *Pinax* surrounded by annotations in the hands of William Sherard and Johann Dillenius.

Left Specimen of sand crocus (*Romulea columnae*) collected by William Sherard in Smyrna, Turkey and given to Sloane. Annotations include the hands of William Sherard and Johann Dillenius.

Sherard was encouraged to start the *Pinax* project by Tournefort. However, for the *Pinax* project to be successful, Sherard needed access to botanical books and herbarium specimens beyond those in his own collections. He had to be able to make decisions about whether polynomials were distinct, or merely synonyms of names already in the *Pinax*. The collections Sloane had at his disposal were essential. However, during the 1720s the Sherard-Sloane relationship soured. Sherard apparently withheld specimens from Sloane, who considered that Sherard preferred continental to English botanists. For his part, Sherard became peeved by what he perceived as Sloane's pettiness in not facilitating essential access to the collection of Leonard Plukenet or even 'Petiver's rubbish'. In 1727, despite his past difficulties with Sloane, Sherard used his considerable influence within the Royal Society to help elect Sloane President of the Royal Society, following Isaac Newton's death.

The *Pinax* project eventually involved James Sherard, Bobart and the first two incumbents of the Sherardian Chair. The project died during the tenure of the second Sherardian Professor, Humphrey Sibthorp. The *Pinax* exists as 130 separate manuscripts, each bound in drying papers used for the preparation of herbarium

Left Specimen of *Stigmaphyllon emarginatum* collected by Hans Sloane in Jamaica and given to William Sherard. The main label is in Sloane's handwriting with annotations by William Sherard and Johann Dillenius.

specimens; some 3,110 bound pages grew to 6,215 pages with the insertion of additional sheets, creating a complex palimpsest. The 16 volumes making up the *Pinax* are an incomplete, unpublished testament to the Sisyphean task Sherard began. With hindsight, the half-century devotion of Oxford-based botanists to Sherard's *Pinax* following his death was a distraction from the botanical developments happening in other parts of Europe.

Sherard's only published book is the pseudonymous SWA's *Schola Botanica* (1689), although he contributed a handful of papers to the *Philosophical Transactions of the Royal Society*. However, as a botanical patron, Sherard had lasting impact. In addition to Dillenius, Sherard's patronage extended to botanists such as John Ray and Pietro Antonio Micheli. Plant collectors, such as Mark Catesby, William Vernon and Thomas More in North America, benefited from his largesse, as they did from that of Sloane. Furthermore, Sherard was instrumental in the posthumous publication of works such as Sebastien Vaillant's *Botanicon Parisiense* (1727) and Hermann's *Musaeum Zeylanicum* (1727) and *Paradisus Batavus* (1698).

Sherard's roles in early-eighteenth-century European botany were as an identifier and nurturer of botanical talent, a maintainer of academic networks and the builder of one of the world's largest pre-Linnaean herbaria. However, the *Pinax*, an intellectual project that occupied the final decades of his life, remained unfinished. Despite their sometimes-difficult relationship, the collections amassed by Sherard and Sloane, the two champions of eighteenth-century natural history in England, benefited tremendously from the mutual exchange of specimens.

Left The impression of a specimen of the cocoa tree (*Theobroma cacao*) on a sheet of drying paper used as the cover of one part of Sherard's *Pinax*.

LEONARD PLUKENET

MARK CARINE AND SUE RYDER

After Leonard Plukenet's death in 1706, most of his natural history collections were purchased by Dr Moore, Bishop of Norwich, before Sir Hans Sloane acquired them in 1710. The herbarium material alone comprised over 8,000 specimens bound into 27 volumes. It included specimens collected from all over the world and was carefully organized and often beautifully presented. Perhaps most significantly, it was extensively catalogued and illustrated by Plukenet in a series of largely self-funded works. Writing on those works, Richard Pulteney suggested that '... no work published before by one man, ever exhibited so great a number of new plants' and Carl Linneaus considered them an 'Opus incomparabile'. Plukenet's herbarium, on which his publications were based, was clearly a very significant addition to Sloane's collection.

Plukenet was baptized in Westminster on 4 January 1642 and attended Westminster School where his contemporaries included William Courten and Robert Uvedale, both of whom also feature in this book. He subsequently matriculated at Hart Hall, Oxford, the precursor of today's Hertford College, although there is no record of him taking an Oxford degree. He was apparently a medical doctor, although where he received his medical qualification is unknown and he does not appear on the rolls of either the Society of Apothecaries or the College of Physicians. He was nevertheless a wealthy individual, leaving 20 houses in Westminster and a farm in the Chilterns when he died. Whether his wealth was inherited or earned through medical practice – official or not – is unclear.

Annotations by Plukenet in his copy of John Ray's *Catalogus plantarum Angliae* (1670) – the standard catalogue of English plants at that time – reveal a thorough understanding of English plants in the field. However, his major contribution was not as a field botanist but rather as a cataloguer of plant diversity in the

Opposite Plukenet's herbarium includes the largest volume in the Sloane collection. The specimens are neatly arranged and carefully labelled with their names in Plukenet's hand. Plukenet used an alphabetical system to arrange his herbarium and so this folio includes plants from several families, including members of the buttercup and pea families.

herbarium, through the critical study of specimens and the botanical literature. His herbarium, though not as extensive or indeed as geographically expansive as those of Sloane or James Petiver, was nevertheless geographically broad in scope and included species and localities not represented in those other major collections. It was largely arranged alphabetically, with specimens labelled with their names and often synonymy and other information. The names of collectors and localities, however, are typically lacking. It is an arrangement that differed from that used in the collections of Sloane and Petiver. The lack of provenance information was unfortunate but it was an arrangement that was well suited to Plukenet's focus on cataloguing species diversity. In conjunction with his herbarium, Plukenet amassed a large library valued at several hundred pounds at the time of his death.

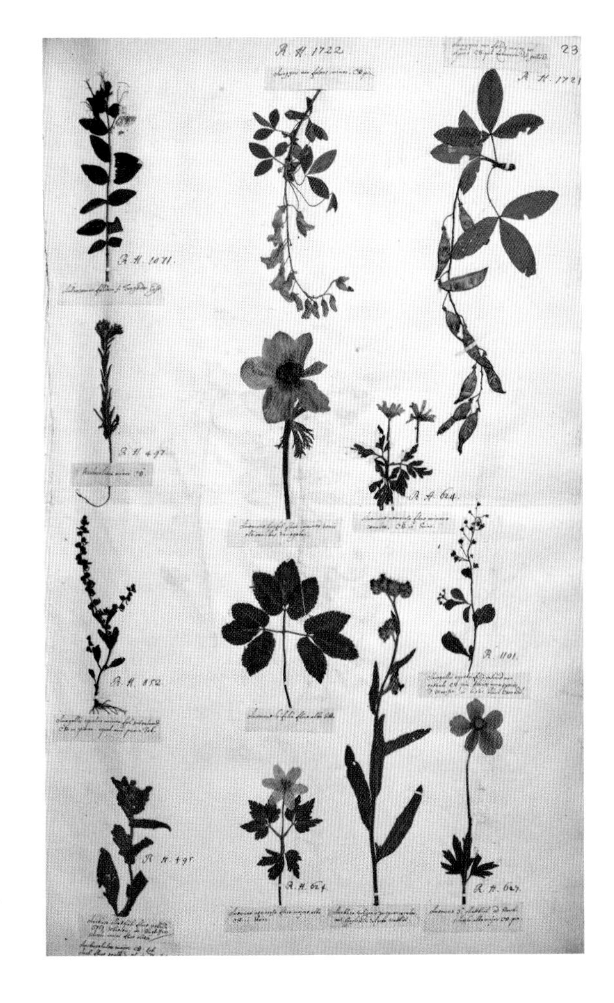

Plukenet's first botanical work was published in 1691 when he was 50 years of age. It was the first volume of his *Phytographia* and it contained 72 plates, each illustrating several plants drawn from herbarium specimens and with names and sometimes brief notes provided at the bottom of each plate. The illustrations were small and, given the limitations of the specimens on which they were based, not all of them included the necessary characters to enable the plants they represented to be identified unequivocally. However, many of the species depicted were new and the eminent naturalist John Ray, in reviewing *Phytographia,* commented that the new species depicted '... may give entertainment to those of the highest form in Botanics, who will here meet with many plants they have not before seen, or it may be heard of'.

Three further volumes of the *Phytographia* followed. The final part was published in 1696 and in the same year Plukenet published his *Almagestium Botanicum,* a catalogue of 6,000 species, arranged alphabetically and cross-referenced to the *Phytographia.* In this and in subsequent works he sought to address problems of synonymy – confusion arising through the use by different authors of different names to refer to the same species – through detailed study of the botanical literature.

Below Plukenet's collection includes species that are today extinct. This illustration from his *Almagesti botanici mantissa* depicts *Trochetiopsis melanoxylon* (bottom right) and *T. erythroxylon* (top left). Both are known only from the island of St Helena. The first is globally extinct and has not been recorded since 1771. The second is extinct in the wild.

Further plates and descriptions followed in his *Almagesti Botanici Mantissa,* published in 1700, and again in 1705 with his final publication, *Amaltheum Botanicum.* In total Plukenet illustrated 2,740 plants and many were based directly on his herbarium specimens. His work was the largest collection of figures of plants then existing and it was frequently quoted by other authors. Carl Linnaeus held Plukenet's works in high esteem and, when he published his great *Species Plantarum* in 1753, the publication that serves as the baseline for the plant names we use today, he described many species using Plukenet's figures. Those figures referred to by Linneaus,

and the specimens on which they were based, consequently remain of particular significance for the accurate naming of plants today.

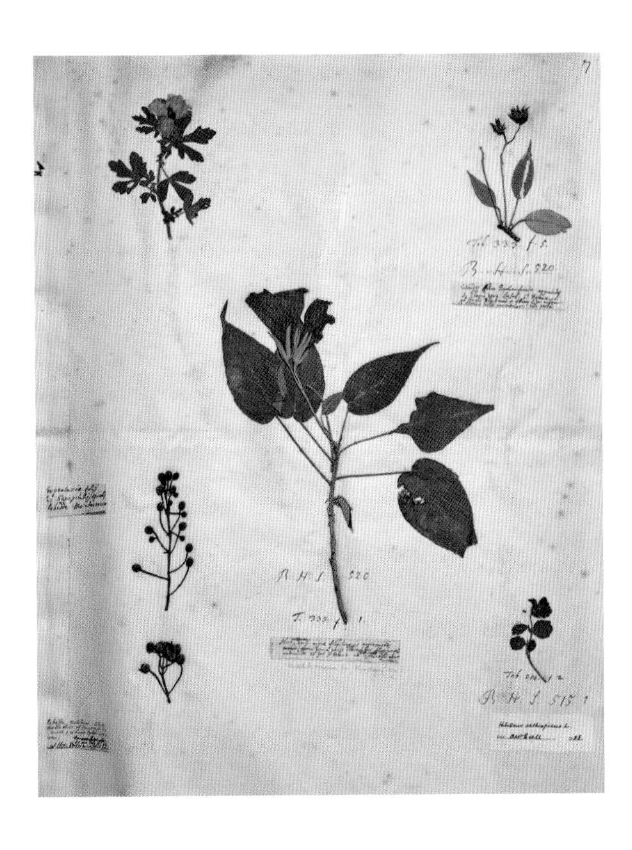

Plukenet's publications and his specimens also provide us with a record of changes in plant diversity and distributions over time. In the *Almagestium Botanicum Mantissa*, for example, he illustrated plants from St Helena, an island in the South Atlantic that has been heavily impacted by humans. He provided the first illustration of the St Helena ebony (*Trochetiopsis melanoxylon*), a species that is today globally extinct and that has not been recorded since 1771. He also illustrated its close relative, the St Helena redwood (*Trochetiopsis erythroxylon*), a species that is now extinct in the wild and survives only in cultivation.

Above A page from Plukenet's herbarium with specimens of *Trochetiopsis melanoylon* (top right) and *T. erythroxylon* (centre) that correspond with the plants illustrated in the figure shown opposite.

Plukenet was honoured by the horticulturally minded Queen Mary who appointed him superintendent of the royal gardens at Hampton Court with the title of 'Royal Professor of Botany' or 'Queen's Botanist'. His contribution to botany was also commemorated by his contemporary, the French botanist Charles Plumier, who named the genus of the Inca nut (*Plukenetia volubilis*) in his honour. However, among his botanical peers in England, his contributions were less celebrated and his relations with them were often strained. For example, whilst John Ray gratefully acknowledged the assistance of Plukenet in the second volume of his *Historia Plantarum* (published in 1693), by 1700 he was writing to Sloane: 'And for Dr Plukenet, I look upon him as an ill-natured man and liable to mistakes, however confident and self conceited he may be…'. A very public disagreement was also played out between Plukenet on the one hand and Sloane and Petiver on the other in

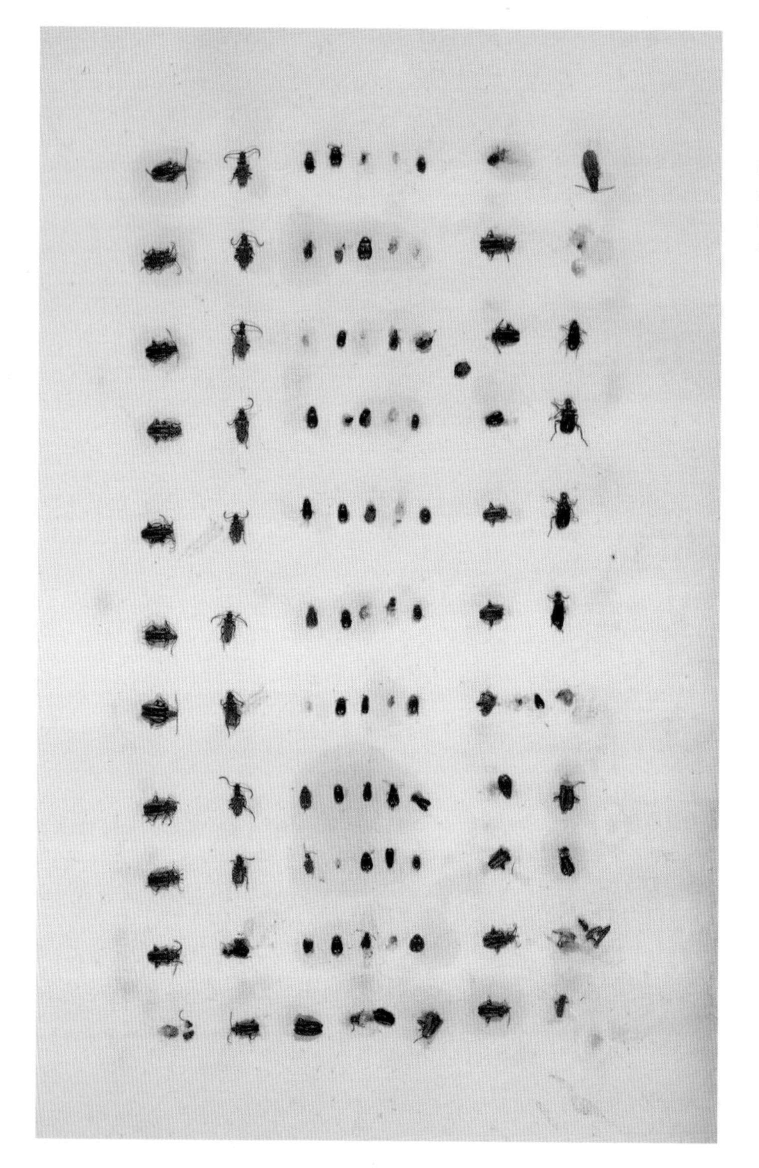

Left A folio of pressed Coleoptera (beetles) in Plukenet's entomological collection.

their writings. It is notable that Plukenet was never elected to the fellowship of the Royal Society, an honour afforded many of his contemporaries. For so prominent a botanist to be overlooked, it may not be coincidental that Sloane was secretary of the Society from 1683.

In later life Plukenet developed an interest in insects and compiled a small collection, which is thought to be the oldest surviving entomological collection. Dating from the late seventeenth century, it comprises nearly 1,500 specimens, all believed to be from the London area. Their method of preservation is distinctive: rather than pinning as entomologists do today, the specimens were pressed on to sheets of parchment in much the same way Plukenet's botanical specimens are preserved. Although many specimens have survived surprisingly well, a good proportion has been lost by becoming detached or damaged due to the unusual preservation method. The specimens have no data but are arranged in an orderly fashion often with specimens of the same species together and largely similar families of insects together, although on some pages the arrangement seems more artistic than scientific. As with his plants, and despite their limited data, Plukenet's insects retain significant, historical, cultural and scientific value. For example, the collection includes species that are now extinct from Britain such as the Black-veined White butterfly (*Aporia crataegi*). This butterfly was first listed for Great Britain in 1667 but has been extinct from the British Isles since 1925 despite efforts to reintroduce the species in 1940 by Sir Winston Churchill.

Plukenet can be seen as a pioneer in both botany and entomology. He assembled a collection to rival those of his contemporaries (or perhaps adversaries) Sloane and Petiver and the body of work based on his specimens remains scientifically relevant today, not least because of its use by Linneaus. Despite their differences, Sloane clearly recognized the significance of Plukenet's collection. After acquiring it following Plukenet's death, he worked on it and catalogued it extensively and its acquisition greatly enhanced the scientific value of his own collection.

Below A portrait of Leonard Plukenet published in his *Phytographia*.

JOHN RAY

CHRIS D. PRESTON

The user of the Sloane Herbarium is constantly aware of John Ray (1627–1705) – more aware of him than of any of the major collectors, more aware of him perhaps than of Sir Hans Sloane himself. This is not because Ray was a major contributor of specimens to the collection. Indeed, there are very few of Ray's specimens to be found in the Sloane Herbarium, all of them in volumes once owned by James Petiver. Rather, it is because the three massive folio volumes of Ray's *Historia Plantarum* were used by Sloane and his assistants as the basic classificatory system to which the plants in the numerous component collections were related. Their work has never been superseded, so that Sloane's annotated, large-paper volumes of *Historia Plantarum* continue in almost daily use as the best available (albeit partial) index to the collection.

Below This engraving, though dating from 1760, was based on a portrait of Ray drawn by William Faithorne in 1690 and commissioned by Hans Sloane.

Unlike many of his contemporaries, such as Petiver and Sloane, Ray was able to make the study of natural history his major preoccupation for most of his life. His interest in plants was first aroused when he was a recently elected Fellow of Trinity College, Cambridge. Having to take a break from his studies in 1650 or 1651, because of illness, he found himself attracted to the wildflowers surrounding him in the fields in spring and 'little by little an interest in botany worked its way into us'. He resolved to 'cultivate Phytology, so neglected & passed over by others' and he published his first botanical work, a catalogue of Cambridgeshire plants, in 1660. He thus stepped into the vacuum left by the death of the aged herbalist John Parkinson in 1650 and the premature deaths of botanist and soldier Thomas Johnson (1644), a casualty of the English Civil War, and William How (1656), a physician who had updated Johnson's work.

Ray left Cambridge in 1662 and then spent three years travelling in Europe, pursuing his natural history and other interests with Cambridge colleagues including Francis Willughby, a kindred spirit who was wealthy enough to help finance Ray's work. On returning to England, Ray was based with Willughby at Middleton Hall, Warwickshire, but he continued to travel, accumulating information for a catalogue of the British flora, *Catalogus plantarum Angliae, et insularum adjacentium* (1670).

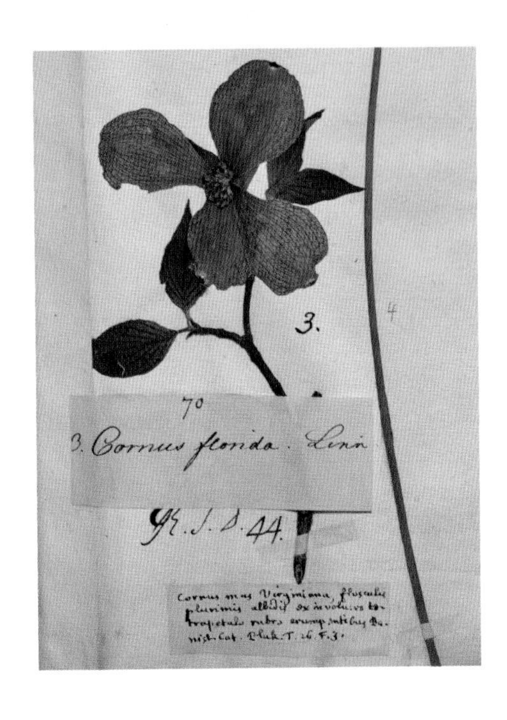

Above A specimen of flowering dogwood (*Cornus florida*) in the Sloane Herbarium, collected in Maryland by David Krieg or William Vernon and identified by Ray in the label at the bottom as a plant previously described by Leonard Plukenet.

The premature death of Willughby in 1672 caused a complete change in Ray's way of life. Although he received an annuity under his friend's will, which allowed him to continue to devote his time to natural history, he stopped travelling and henceforth lived a sedentary life. In 1679 he returned to his home village of Black Notley in Essex, where he lived with his wife Margaret and their daughters. In the immediate aftermath of Willughby's death Ray concentrated on finishing his friend's incomplete works on birds, *Ornithologiae libri tres* (1676), and on fish, *De historia piscium* (eventually published in 1686). It was only in the 1680s that he began to concentrate again on plants, publishing a new scheme of classification (*Methodus plantarum nova*, 1682) and then *Historia Plantarum*, an account of all the known plants in two folio volumes (1686, 1688), each with some 1,000 pages.

Ray was a deeply religious, serious and rather abstemious figure, chided for his 'utter moderation' by his friend and fellow naturalist Martin Lister, who had noticed his 'extremely sparing and careful' diet when they were eating together in Montpellier. Neat, organized and meticulous, and so dedicated to natural history that he might these days be regarded as a workaholic, he published some 20 books on botany, zoology, philology and theology between 1660 and 1705. He was by nature cautious, self-depreciating and a natural collaborator who corresponded with a wide range of botanists in Britain and overseas, including many whose specimens eventually ended up in the Sloane Herbarium. Elegant tributes to his friends and collaborators are scattered throughout his works.

Despite the great disparity in their material circumstances, there was clearly some similarity between Ray's character and that of the equally industrious Sloane, who was described by Thomas Birch as 'extremely temperate both in eating & drinking'. Their correspondence began in the mid-1680s when Ray's thoughts were 'almost wholly employed at present in the carrying on the *history of Plants*'. Sloane supported Ray's work on this, making some suggestions about its contents and offering to do all he could to help him obtain any material he needed from Paris or Montpellier; Ray later thanked him for 'the many informations and advices you have given me'. However, it was not until Ray was preparing the third (supplementary) volume of *Historia Plantarum*, which was eventually published in 1704, that he worked closely with Sloane. The letters they exchanged give a vivid impression of the rate at which plants were being discovered in these years. In August 1701, for example, at least three years after Ray had first completed the text of his supplement, he was still adding species to it, working on 'two great volumes of dried plants' lent to him by Sloane, from Aleppo and the Straits of Magellan (the latter collected by George Handisyd), and telling Sloane that he would 'be glad to see your China-plants' (collected by James Cuninghame). Two months later, in October 1701, he reported that Dr William Sherard 'hath lately sent me a parcell of about 230 dried plants, received from the Prince

Opposite Ray's account of the lady's slipper orchids in the third volume of *Historia Plantarum* (1704). He noted of no. 14 (*Cypripedium acaule*) that the colour of the dry specimen was not apparent, but it seemed to have been yellow; in fact it is usually magenta or pink.

HS. 328 – 8 – Helleb. seu Calceol. marian. hirsutior flore maximo purpurascente H. Ox.
H. S. 37. 48. Helleborine vernalis fl. purpurascente Mariana.
H. S. 37. 37. Helleborine Floridana, biflora purpureo fl. caule & foliorum nervis hirsutis pubescentibus. Pluk. Amalt. T. CCCCXVIII.

3. Helleborine *Virginiana* seu Calceolus angustifolius flore amplo purpureo D. *Banister Pluk. Almag. Bot.*

(H.S. 312. 127) 4. Helleborine Alpina lutea seu Calceolus *Mariæ Icon. Robert. Virginiana*, flore rotundo luteo D. *Banister Pluk. Almag. Bot.*

5. Helleborine similis *Canariensis*. Flos Spiritûs sancti vulgo & *P. B. P. Pluk. Almag. Bot.*

6. Helleborine *Mariana*, Bupleuri angustissimo folio, purpurascente flore, caule aphyllo: *Pluk.* Ab hac (inquit) parùm distat Helleborine angustifolia, flore rubente *Virginiana Banisteri.*

Triphora trianthophora (Sw.) Rydb.

7. Helleborine *Mariana*, flore pallidè purpureo, trianthophoros *Pluk. Mantiss.*

92. 85. H. S. 93 – 198 – 242. 15. 8. Helleborine *Mariana* monanthes, flore longo purpurascente Liliaceo *Ejusd. ibid.*

312. 130 H. S. 93 – 198. – 74. 85. 9. Helleborine affinis planta *Mariana*, Herbæ Paridis facie, quinquefoliata *Ejusd. ibid.* vid. Herba Paris. 351.

H. S. 312. 128 – H. S. 92. 84. 10. Helleborine *Virginiana*, flo. rotundo magno ex purpureo albicante *Banist. Cat. Mss. Pluk. Mantiss.*

Cypripedium acaule Aiton

H. S. 93 – 197 – 74. 87. | 242. 14 | 312. 127. 11. Helleborine, Calceolus dicta, *Mariana*, foliis binis è radice ex adverso prodeuntibus, flore purpureo *Ejusd. ibid.* idem c. Carolin. bifol. lateralibus florum petalis praelongis admodum & angustissimis. Amalth. 115. T. 418. f. 1.

H. S. 93 – 197 – 12. Helleborine Calceolus dicta, *Mariana*, flore gemello candido, venis purpureis striato *Ejusd. ibid.* Binis quoque foliis (inquit) hæc prædita est, alterum ad radicem, alterum ad caulem situ alterno promens, cujus florum petala sunt multò latiora & breviora cæteris hujus sortis. Æstivo tempore floret.

Cypripedium calceolus L. var. parviflorum (Salisb.) Fernald

Cypripedium acaule Aiton

13. Helleborine flore rotundo, seu Calceolus *Americanus* maximus bifolius. Calceolus Marianus *Canadensis Cornut.*

Flos huic maximus est, colore albo rubris lineis striato, describente *Cornuto*: nam in siccâ nobis visâ, quam ex *Marilandia* redux secum attulit D. *Vernon*, color non comparuit. *Folia* maxima & latissima, bina tantùm è radice egressa, in caule nulla.
Non tantùm in *Canada*, sed & in *Virginia* & in *Marilandia* invenitur.

C. acaule Aiton

14. Helleborine flore rotundo seu Calceolus *Marilandicus*, caule foliis latis nervosis glabris, quatuor aut pluribus, cincto.

Caulis huic semipedalis aut altior, parùm lanuginosus, foliis 4 latis & subrotundis, in apicem tamen acutum terminatis, basi sua caulem amplectentibus, nervosis, nervis arcuatis circumferentiæ propemodum parallelos ductis, vestitus. *Folium* è cujus sinu flos exit majus est & latius quàm in aliis speciebus, Flos ipse minor quàm præcedentis. Color in sicca non apparet, videtur tamen luteus fuisse.
In *Marilandia Americæ* provincia collegit D. *Vernon.*

15. Helleborine flore rotundo seu Calceolus *Marilandicus*, foliis longis acuminatis glabris, 5 aut pluribus in caule.

C. calceolus L.

Caule est semipedali aut altiore, quinque [in planta nobis visa & descripta] *foliis* cincto angustioribus quàm præcedentium duorum, satis tamen latis, in longos & acutos mucrones productis, 7 nervis insignioribus striatis. *Flos* in summo caule unicus, ut in sicca videtur luteus, è sinu folii angustioris & acuti exit, pediculo longiusculo. Petala floris lateralia oblonga & angusta sunt, erectum etiam & dependens longiora, angustiora & acutiora sunt quàm in reliquis speciebus.

16. Helleborine flore rotundo seu Calceolus *Marilandicus*, foliis angustioribus acuminatis circum oras lanuginosis tribus quatuórve in caule.

C. reginae Aiton

Altitudo huic semipedalis. *Folia* angustiora quàm primæ & secundæ, in longos & acutos mucrones producta, lanuginosa, lanugine circa margines valde conspicua, nervosa adeò, ut tota ex nervis contexta videantur, insigniores autem nervi, aversa parte præcipuè conspicui, 3, 4ve ex utroque latere costæ mediæ decurrunt. Floris labellum, calopodium imitans, brevius & tumidius pro magnitudinis ratione quàm in reliquis, petala autem lateralia breviora & latiora sunt.
In eadem provincia observavit & collegit D. *Vernon.*

Cypripedium reginae Aiton

17. Helleborine Calceolus dicta, *Mariana*, Lilii convallium folio unico, caule tomentoso, floris utriculo maximè elongato *Pluk. Mantiss.*

18. Hel-

HS. 89. 37 | 242. 14. ead. c. 11ma Helleborine Calceolus dicta Carolinensis bifolia lateralibus fl. petalis praelongis admodum & angustissimis. Pluk. Am. 114. T. 418. f. 1.

HS. 264. 93. Calceolus Mariae fol. alternis petalis longioribus acutis.

Left Specimens of pink lady's slipper (*Cypripedium acaule*) (top) and yellow lady's slipper, (*C. parviflorum* var. *pubescens*) (bottom) collected by William Vernon in Maryland and now in the Sloane Herbarium.

of Catholica, most of them new & unknown to me'; these came from the Prince's garden near Palermo, Sicily. In the end the supplementary volume of *Historia Plantarum* extended to over 950 pages.

Ray faced some practical difficulties in his working relationship with Sloane. Worries about the safe carriage of parcels between Black Notley and London, and the negligence of the postboys, recur repeatedly in their correspondence. Ray's small cottage at Black Notley was not an ideal working environment. At an early stage in their correspondence (11 February 1685) Ray had to explain to Sloane that a fungus he had been sent in a letter 'unluckily slipt out ... & it being candle-light & company in the room, was trode to pieces'. Later (2 June 1699) he had to 'beg your pardon for having in some measure defaced' books lent to him by Sloane, 'partly by a childs unlucky scattering ink upon them'.

The major difficulties Ray faced in dealing with Sloane's plants were not, however, caused by the cramped accommodation at Black Notley but with his fundamental problem in studying dried specimens. In his years of travel he based his knowledge of species on the description of living plants in the field and, initially, their cultivation in a small garden in Cambridge. He agreed with Sloane (17 November 1685) 'that those who cultivate plants, & have the liberty & freedome to pluck up & observe their roots, have a great advantage of those who see them only in one state'. In Black Notley, however, his garden occupied 'a cold soil, & ill situated place'. The dried specimens he received 'cannot represent all the principall parts' and 'a man that relies wholly on dried specimens ... must needs commit many mistakes'. Thus, though he never ceased to ask Sloane to send specimens for him to examine, he was frequently unable to make much of them. He 'was very much taken with the beauty' of the 'large & fair samples of rare & non-descript plants' collected by William Vernon and David Krieg in Maryland and sent to him by Sloane in 1699, but fruits and seeds were 'scarce to be seen at least perfectly discerned in any of them, neither the colour or figure of the flower, without marring the specimens ... the stature to be known but in few, & nothing of the root'. Nevertheless, there are many examples in the Sloane Herbarium of specimens annotated by Ray in the course of his work on this book. Many species in Sloane's herbarium were well-known and could be matched with those described earlier. Despite his difficulties, Ray was able to make use of others in preparing the species accounts in *Historia Plantarum*, such as the species of lady's slipper orchids (*Cypripedium*) he saw from Maryland. However, many could not be described adequately; Ray told Sloane that Handisyd's 'Magellanick Plants were of little use to me' and although some of these collections have subsequently been studied by scholars, other plants still lie unidentified in the Sloane Herbarium.

Ray struggled with poor health throughout the 1690s, describing his symptoms at length to Sloane in his letters and receiving his advice. In January 1705 his health finally gave way, and as he lay dying it was to 'the best of friends', Sloane, that he chose to send his last letter. 'God requite your kindnesse expressed any ways towards me an hundred fold, blesse you with a confluence of all good things in this world, & eternal life and hapinesse heer after & grant us a happy meeting in Heaven'. Even on his deathbed Ray might not necessarily have considered that his work was over, for he had written earlier that 'it may be ... part of our business and Employment in Eternity, to contemplate the Works of God'.

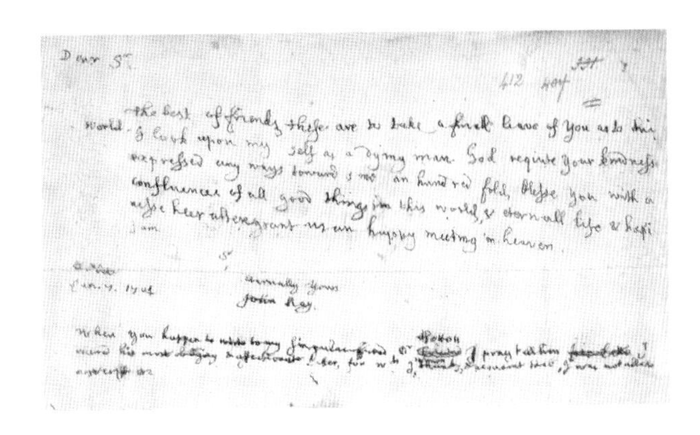

Dear Sr

the best of friends these are to take a final leave of you as to this world. I look upon my self as a dying man. God requite your kindnesse expressed any ways towards me an hundred fold, blesse you with a confluence of all good things in this world, & eternall life & hapinesse heer after & grant us an happy meeting in heaven.

I am Sr

eternally yours
John Ray.

When you happen to write to my singular friend Dr [illegible] I pray tell him [illegible]

Below John Ray's last letter to Hans Sloane, written on 7 January 1705 (new style calendar).

THE GARDENERS

MARK CARINE AND ROBERT HUXLEY

Sloane was not a gardener but throughout his long life he maintained a close association with gardens and with gardeners. As a young man, gardens were where he learnt his botany; and in later life, through his patronage, he would play an important role in their development. Sloane's herbarium benefited greatly from this association and gardens and gardeners were a rich source of specimens.

When Sloane left his native Ireland at 19 to train as an apothecary in London, botany was an integral part of his training. The Chelsea Physic Garden, established by the Worshipful Society of Apothecaries in 1673, was where his botanical skills were honed. In 1683, after completing four years of training, Sloane travelled to continental Europe where gardens again featured prominently as he sought to further develop his botanical expertise. In Paris he continued his botanical studies at the Jardin du Roi under the eminent professor of botany Joseph Pitton de Tournefort. He then spent several months in Montpellier where he studied medicinal plants in the city's botanical garden before being awarded a medical degree at the University of Orange in 1683. Sloane made herbarium specimens during this time that were collected in the gardens around London, Paris and Montpellier. These specimens are typically small, with several mounted on a sheet and with the glue sometimes rather carelessly applied. They look like specimens made by a man learning his craft and contrast markedly with the specimens Sloane would make later.

Later in life Sloane, the wealthy society physician and Jamaican plantation owner, acquired the Manor of Chelsea and with it, he

Opposite Ornamental flowers from the herbarium of George London.

became landlord of the Chelsea Physic Garden where he had studied years earlier. In 1722 he secured the future of the garden, granting the Society of Apothecaries control in perpetuity for an annual payment of £5. Sloane's Deed of Covenant also required that 'ffifty specimens or samples of distinct plants well dryed and preserved and which grew in the said garden the same year with their respective names' were supplied to the Royal Society every year 'ffor improving natural knowledge'. It is clear from this that Sloane recognized the scientific value of gardens and rightly so; late seventeenth- and early eighteenth-century gardens were places of experimentation, innovation and discovery. In 1716 the celebrated Hoxton nurseryman, Thomas Fairchild, created the first artificial plant hybrid. The Sloane Herbarium contains a specimen of 'Fairchild's mule' as it would come to be known, a cross between a sweet william and a carnation pink. Fairchild's experiments with hybrids helped to establish the role of sex in plants and would revolutionize the world of horticulture. At the same time, technological advances were making possible the cultivation of a much wider array of species. For example the first stove house in England was completed at Chelsea Physic Garden in 1681, allowing tender plants, that would otherwise perish, to survive the cold English winters. And new plants were arriving in English gardens from across the world at a remarkable rate, in part through plant-hunting expeditions that Sloane and others helped to finance.

Above A statue of Sir Hans Sloane at Chelsea Physic Garden. Sloane granted the garden to the Society of Apothecaries in perpetuity after acquiring the Manor of Chelsea.

Gardens and gardeners from across Britain and Europe and also from further afield are represented in the herbarium. Here we look at four of them. First, Sloane's friend and neighbour Mary Somerset, the Duchess of Beaufort, who was one of the most passionate plant collectors of the seventeenth century. Sloane acquired her herbarium, probably after her death in 1715. Arranged in 12 large volumes, it provides a remarkable record of the diversity of plants cultivated in her gardens and of the species that she introduced.

Robert Uvedale's large herbarium was also acquired by Sloane. A schoolmaster and clergyman by profession, Uvedale was also a renowned horticulturalist with a garden in Enfield and a reputation for his skill in cultivating exotics. The 14 volumes of Uvedale's herbarium are beautifully arranged and carefully classified and certainly demonstrate his skills as both a gardener and a botanist. George London has been considered one of the most influential nurserymen and garden designers of the late seventeenth and early eighteenth centuries. His contribution to the Sloane Herbarium was more modest but it is one that again shows great botanical awareness and skill, embracing the latest scholarly work in its arrangement.

Finally, Philip Miller, who perhaps demonstrates more than any other contributor the role of gardens as a source of specimens for documenting and classifying plant diversity. Sloane was instrumental in Miller's appointment at Chelsea Physic Garden in 1722 and Miller, in turn, contributed many volumes of plants to the Sloane Herbarium. Under Miller's leadership, Chelsea would become a leading botanic garden and he was responsible for the introduction of many species into cultivation in Europe. Specimens of those plants provided new additions to Sloane's herbarium and indeed, many were entirely new to science.

Below A specimen of 'Fairchild's Mule', the first documented artificial hybrid, that was made by Thomas Fairchild who crossed a sweet william and a carnation pink.

MARY SOMERSET, DUCHESS OF BEAUFORT

VICTORIA PICKERING

Mary Somerset, Duchess of Beaufort, was the eldest daughter of Arthur Capel, first Baron Capel of Hadham. After her first husband died, she married Henry Somerset in 1657. He was made the Duke of Beaufort in 1682, and in 1684 they inherited the Badminton Estate in Gloucestershire from his cousin. They would later purchase Beaufort House in Chelsea. At both properties Mary, who had considerable wealth and a place amongst the aristocracy, made impressive investments in the gardens. She was considered highly by all sorts of people for her horticultural expertise and was regarded as 'a Patroness, who dayly makes appeare the transcendent wealth of the Vegetable Kingdome'. Jacob Bobart, who had been in charge of the physic garden at Oxford, wrote this remark in a letter to the Duchess in 1694. It would be just one of the many instances of praise given to the 'success and prosperitie of [her] glorious Gardens'. Ultimately, these were spaces in which Somerset could expend both money and time to create well-known plant collections, and through which she could develop relationships with men such as Sir Hans Sloane.

Below An engraving of Mary Somerset by the English engraver Joseph Nutting, thought to have been completed between 1690 and 1722.

The Sloane Herbarium houses 12 beautiful herbarium volumes that were created by Somerset. Sloane incorporated them all into his own collection, much as he did with those of other collectors. In addition, Sloane's botanical collection of Vegetable Substances includes upwards of 400 botanical items listed from Somerset, making her one of the most significant contributors to Sloane's botanical collection.

Somerset collected and cultivated plants from around the world. These practices of plant collecting were truly wide reaching in terms of geographical scope not just in terms of plants, but also people. Today we find Somerset's surviving correspondence and careful record-keeping held at the British Library and the Badminton Estate. When considered alongside her surviving plant specimens, it becomes clear that she was in correspondence with all sorts of people around the world. Her role in the movement and exchange of botanical knowledge during this period is not to be ignored.

As a woman, Somerset could not be elected to fellowship of the Royal Society but this did not hinder her ability to make strong connections to leading natural historians, philosophers, botanists, horticulturalists and Royal Society Fellows. She was regularly engaged in the exchange of ideas and plants with men such as Jacob Bobart, superintendent of the Oxford Physic Garden and with others featured in this book such as Hans Sloane, James Petiver and William Sherard. These were domestic and global interactions. In 1694, for example, Bobart informed Somerset that 'I send now a packet of such seeds as to me seem hopefull, partly East Indians, partly West Indians, and perhaps some out of our Garden'. George London, who had been William III's gardener at Hampton Court, similarly supplied Somerset with exotic plant specimens, this time sourced in New England and Virginia as well as in the West Indies.

The specimens now in Sloane's collection show that many individuals sent Somerset exotic plant material. For example, the naturalist and Royal Society Fellow Richard Bradley, who published several works on natural history, sent her material from Holland in 1714 that had been collected in the Cape of Good Hope, while 'Spainish merchants' and a 'merchant in Genoa' ensured she received 'Indian cane'. Somerset also received 'Seeds taken out of a ship from the East Indies', 'Sugar beans' and 'Sea side beans from Carolina', as well as other specimens from Fort St George in India.

Somerset was accomplished both in terms of cultivating impressive gardens and forming her own collection of dry-pressed plant specimens. These specimens reached Sloane via two main routes, the first as direct gifts and exchanges with Sloane while Somerset was alive. This was part of an important relationship in which Sloane and Somerset were friends. Secondly, Sloane appears to have been bequeathed Somerset's 12-volume herbarium after her death in 1715. Sloane thought extremely highly of Somerset's horticultural abilities. He professed to this in the preface to his *Voyage to Jamaica* (1707):

> *but especially at Badminton in Gloucester-Shire, where they are not only rais'd some few handfuls high, but come to Perfection, flower and produce their ripe Fruits, even to my Admiration; and that, by the Direction of her Grace the Duchess of Beaufort, who at her leisure Hours, from her more serious Affairs, has taken pleasure to command the raising of Plants in her Garden.*

Below A fine specimen of prickly pear cactus (*Opuntia*) in the Duchess's herbarium. As both spiny and succulent, cacti are notoriously difficult to dry and prepare as herbarium specimens.

It is clear from the numerous letters written by Somerset to Sloane that they grew to be great friends, and that this, in part, sustained the correspondence between them. Somerset regularly signed her letters to Sloane as 'Y^{r} affe^{ct} friend', and no doubt they would have seen each other when they were both residing in London, Sloane in Bloomsbury Place and Somerset in Chelsea. Indeed, in one undated letter to Sloane, Somerset was very insistent that she see him and wrote, 'If bussines does not earnestly require y^{r} being in towne all day, I should go my journie much more quietly, if I could speake [with] you once more before I go, wee shall be at home al day.' Sloane also advised Somerset on personal and family matters and acted as her and her family's physician. In the 1690s, Somerset wrote to Sloane: 'Tho I have this day taken phisick, I have something to aske y^r assistance in, that I am not willing to employ any body else to do.' She goes on to request that Sloane help her in obtaining the services of the botanist William Sherard to 'bee wth him [her grandson] as a companion'. Indeed, this appeal was successful. It was only three months later when she wrote to Sloane saying, 'I thanke you for the assistance you gave mee in procuring Dr Sherwards being with my Grandsonne. I am very much satisfied wth him, & that which is much better, the young Lord is so too.'

Above Sloane gave seeds to the Duchess that she grew in her gardens. Here we see a plant originally sent to Sloane from China which has been both pressed and dried by Somerset, and preserved in a small box by Sloane in his collection of Vegetable Substances.

Similarly, many of Somerset's letters to Sloane involve her and her family's health. She even told him, 'but not att all to flatter you I must tell you the same I do to every body else, that I have so much confidence in y^r care of [my daughter], (besides y^r skill) that I thinke her as safe as if I were there myselfe to take care of her'. In fact, this example demonstrates how the relationships Sloane had with his medical patrons were important in allowing him to form his botanical collection.

Intermixed with intimate discussions about health and family are references to gardening, scholarly works and the collecting, cataloguing, planting and raising of plants. In a letter sent in the late 1690s, Somerset described 'a small box with some fruits that had ripened in the warm places' at Badminton which meant that she had to send them to Sloane sooner than she had intended. She also informed him that she had sent 'a small parcel of Badminton plants (all except a very few) of my owne raising, I am sorry I did not make the booke bigger'. Somerset regularly discussed her success or lack of success in growing plants from seed and wrote

from Badminton about the thriving 'silke cotton' and a 'Gourd', along with the plants she raised from the 'Shaddock' thanks to the hot summers that 'brings [her] plants to such a height'. Somerset and Sloane also discussed the exchange of scholarly and scientific works and Somerset sent Sloane 'many thanks' for 'the trouble of getting mee a booke bound for my Parchments', which she would go on to use to catalogue the plants she had raised. This was not just a one-way exchange of specimens. Sloane appears to have had a vested interest in sending Somerset material. He not only sent her books, he also sent her seeds, sometimes only a few and at other times, many more. Somerset went on to cultivate them and preserve dried specimens of the plants.

Above The Duchess of Beaufort was celebrated for her cultivation of plants such as auricularias (*Primula auricula*) which are shown here, carefully preserved and labelled in her collection.

Somerset's gardening practices, her expertise in cultivating plants and her connections with the Royal Society and people around the world are well known. Her contributions to Sloane's collection also demonstrate the considerable results that she achieved with her plants and the high esteem in which people like Sloane held her. While she was known for growing flowers such as carnations, tulips and auriculas, Somerset was also responsible for the introduction of a significant number of plants into horticulture in England.

Living not too far from each other in London, Somerset and Sloane no doubt had numerous meetings, conversations and exchanges involving plant material and botanical discussions that are not documented anywhere. The formation of Sloane's herbarium was helped, in part, by particular sorts of social relationships that went beyond the simple exchange of botanical objects. Here, it was through friendship with an expert cultivator.

Left The Duchess cultivated a wide array of plants but succulents, including this *Aloe*, were a particular focus. This specimen once again highlights her impressive skill of pressing, drying and preserving plants.

ROBERT UVEDALE

VICTORIA PICKERING

Robert Uvedale was born in London and met Leonard Plukenet and William Courten, both featured in this book, during his education at Westminster School. He went on to Trinity College, Cambridge, and after graduating in 1663, was appointed master at the Free School in Enfield in 1664. From 1676, he focused his attention on a nearby school that he had opened called the Palace School. Uvedale, a teacher of great repute, taught sons of noblemen and gentlemen including two of Sir Hans Sloane's nephews. School education was not Uvedale's only connection with practitioners of natural history, though. His talent for cultivating plants and his impressive garden in Enfield were widely known at the time. He was a regular correspondent with Sloane and other prominent naturalists of the period such as William Sherard and the Yorkshire physician Richard Richardson. Sloane acquired Uvedale's 13-volume herbarium after Uvedale's widow died in 1740. Uvedale also actively contributed a significant number of specimens to Sloane's Vegetable Substances collection during his lifetime, the aesthetically unusual collection comprised of over 12,500 botanical specimens such as seeds and beans that Sloane had placed into small glass boxes and sealed with ornate paper.

Just as coffee-houses and the meeting rooms of the Royal Society had provided the ideal space for both science and socializing to take place, gardens were also disposed to this. Gardens were important social locations for both private and public purposes. In fact, the Royal Society urged its members to become authors on horticultural and agricultural matters, and fellows such as John Evelyn and James Petiver were influential in improving gardening techniques, swapping and importing plants, expanding classification systems, changing taste and promoting horticulture as both an art form and a science.

Left Specimens of parrot tulips found in Robert Uvedale's collection. These flowers would have been highly prized and expensive in the eighteenth century.

James Petiver presented notable accounts to the Royal Society of plants growing in different gardens around Britain (and Europe). In his *Account of Divers Rare Plants Observed This Summer, A.D. 1713 in Several Curious Gardens about London* Petiver mentions a number of plants, such as a grass (*Gramen typhinum*) that had been 'this Summer in Chelsea Garden raised from Seed'. In the same publication Petiver highlights the significant gardening practices of 'Dr. Uvedale at Enfield' alongside other gardens of note. Sloane similarly regarded the skill and expertise of this horticultural expert highly. At the end of the preface to Volume One of his *Voyage to Jamaica* (published in 1707), Sloane notes the importance of gathering plants in the West Indies and the sorts of people across Britain and Europe who were cultivating them in different types of gardens:

> *The Plants themselves have been likewise brought over, planted, and throve very well at Moyra, in Ireland, by the Direction of Sir Arthur Rawdon; as also by the Order of the Right Reverend Dr. Henry Compton, Bishop of London, at Fulham; at Chelsea by Mr. Doudy; and Enfield by the Reverend Dr. Robert Uvedale; and in the Botanic Gardens of Amsterdam, Leyden, Leipsick, Upsal, &c.*

Sloane sent these different 'gardeners', including Uvedale, samples of seeds. Uvedale was not a Fellow of the Royal Society and yet although trust was more likely accorded to someone because of their association with such an established institution, Uvedale remained a prominent individual in early eighteenth century botanical exchange. As is clear from the words of Petiver and Sloane, a lack of Royal Society fellowship did not impinge on how other naturalists viewed Uvedale's impressive garden in Enfield. Indeed, considered to be a 'gentleman gardener', Uvedale had a wide circle of friends and acquaintances who shared similar interests in natural history and horticulture. His correspondence with the physician Richard Richardson shows that he regularly exchanged information, books and plants with influential naturalists including several featured in this book: Edward Lhwyd, James Petiver, John Ray, Leonard Plukenet, William Sherard and, of course, Sir Hans Sloane. Uvedale's gardening practices and botanical knowledge were held in high regard by many people.

Uvedale's herbarium has been described as one of the best preserved in the Sloane Herbarium. The volumes have been organized according to Ray's classification and it has also undergone some rearrangement while being absorbed into Sloane's collection.

For example, the specimens have been mounted onto larger sheets. In 1718, Uvedale described to Richardson that he had furnished his herbarium with plants from his own garden and that William Sherard had helped him in 'correcting' his *hortus siccus*. Today the herbarium contains various fine specimens of cultivated plants as well as British and North American orchids, and also seaweeds and other plants that Uvedale notes were 'brought me from Barbadoes 1721 by the ingenious Mdm Titus'.

Various British collectors of the time contributed to Uvedale's herbarium: James Sherard collected the first British specimen of *Melampyrum arvense* L., commonly known as field cow-wheat, in Norfolk and a variety of plants from Wales by way of Richardson are found scattered throughout the volumes. From much farther afield are specimens from Fort St George, Bengal and Pegu and it quickly becomes clear that some of the collectors and correspondents mentioned among these volumes are not found elsewhere in the Sloane Herbarium.

Uvedale also contributed specimens to Sloane's botanical collection of Vegetable Substances that were sent from the East Indies (a broadly defined geographical region in the eighteenth century). He had received 100 specimens from Fort St George (Madras) in India as well as 200 items from 'Siam' (Thailand). A further 50 of these samples came into Uvedale's possession via his brother, Henry Uvedale, who was an East India Company captain on board the *Herbert* that sailed to Bengal in 1682 as well as to Benkulen in Sumatra in 1685.

Uvedale never travelled to India or Thailand to collect specimens, he remained in North London. However, he was both receiving seeds to cultivate in his garden at Enfield from well-known naturalists of the period and acting as a conduit for supplying seeds to other collectors.

Above Uvedale had a far-reaching botanic network and his herbarium includes many specimens received from collectors overseas. Among those were plants such as the one shown here sent by 'the ingenious [Madam] Titus' in Barbados.

Uvedale exchanged letters, books, news, specimens and seeds with a variety of people across Britain. He received samples from correspondents in Britain, such as 'northern plants' from Richardson and seeds from much farther afield, and he regularly moved this material onwards to his correspondents. This is seen particularly clearly when he writes to Richardson in 1699 that he had received a 'small parcel sent by Dr. Hotton' – that is, Peter Hotton, curator of the Leyden Botanic Garden – who himself had acquired 'them by chance from Denmarke, one of their ships last year touching att the Cape [of Good Hope]'. Uvedale had been sent some seeds from William Sherard from Rome of which 'there are some very good plants among them' as well as 'A few I have received from the Oxford garden, and a pretty numerous parcel from a Scotch gentlemen'.

In some instances, Uvedale directly appealed to Sloane, as a friend, for seeds. In 1698, for example, various problems had caused Uvedale to be without plants including the death of one collector named as 'Dr Carr' in Lyon, France, while a ship from Carolina had

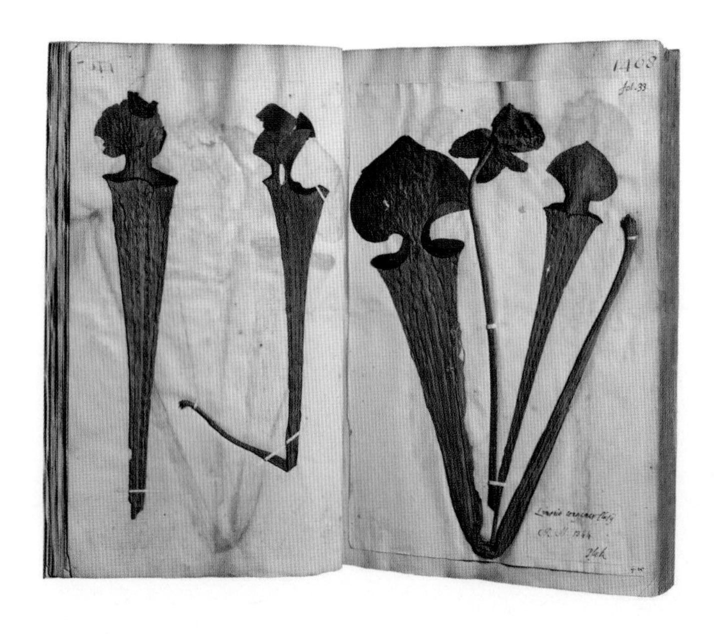

Left The plants preserved in Uvedale's herbarium are beautifully pressed and arranged. This double-folio spread shows the unusual shape of the pitcher plant (*Sarracenia*) and its leaves.

been 'wrect on [the] Isle of Wight'. This meant that Uvedale required help from his 'Friends' in procuring seeds. But as the numerous letters exchanged show, Sloane sent seeds to Uvedale on an ongoing basis as well. Some of these plants would come back into Sloane's possession in the form of herbarium specimens when Sloane acquired Uvedale's herbarium in the 1740s.

Uvedale could also be found regularly updating his botanic friends with news of his own (ill) health, especially when it affected his garden. In 1701 he wrote that he had 'injoyed so little health this spring' that he had 'been forc^{d} almost wholy to neglect [his] garden'. By 1720 Uvedale's health had become a hindrance to his successful cultivation of plants and as a result he turned his attention to his herbarium, and to taking 'a little pleasure in turning over [his] Hortus Siccus'.

Above Almost 300 years since the plants were originally collected, these specimens of spring bulbs from Uvedale's collection highlight the vibrancy of the colours still preserved today.

Uvedale was entwined within a regular exchange of letters around Britain that extended to Europe and farther afield and revolved around the business of botanical exchange, the pleasure and profits of gardens, methods of cultivating different plants, the improvement of garden stocks, as well as natural history more broadly. It was a correspondence sustained through expressions of friendship and affection.

Uvedale was known for the size and range of his plant collection, and the cedar of Lebanon in his garden may well have been the first one to be introduced into the country. He influenced the plant world in other ways, too. James Petiver named a new genus, *Uvedalia*, after him. Phillip Miller retained this in his *Gardeners Dictionary* and it lives on in the Linnaean binomial system as the species epithet in the name of a North American daisy, *Polymnia uvedalia*.

PHILIP MILLER

JACEK WAJER

Tucked away in the southeastern corner of London's affluent Royal Borough of Kensington and Chelsea lies one of the most important sites in the history of British botany and horticulture, the small but mighty Chelsea Physic Garden. The Worshipful Society of Apothecaries founded the garden in 1673 as a place where its members could learn how to grow and identify medicinal plants. Before long this small plot of land by the River Thames developed into an important centre for the introduction of hundreds of new species of ornamental plants from around the world into British and European horticulture. It was here that Philip Miller, one of the most influential gardeners of the eighteenth century, reigned as a director for nearly 50 years from 1722 to 1770. Under his leadership the Chelsea Physic Garden was transformed into one of the most famous botanical gardens of Europe, renowned particularly for its collection of North American plants. Thanks to his triumphs at Chelsea, Miller became a highly sought-after gardening adviser, a favourite amongst the British aristocracy, and a prolific author of horticultural manuals. His most famous work, *The Gardeners Dictionary*, a practical guide to plant cultivation, was published between 1731 and 1768. Much of Miller's success was owed to Sir Hans Sloane, who recognized his talent at just the right time and propelled his career onto the path of horticultural stardom.

Miller's passion for plants was in his blood. His Scottish father owned a market garden in Deptford, southeast London, where young Philip had cut his horticultural teeth. By the age of 30 he was running his own plant nursery in St George's Fields in Southwark, but high costs of rent forced him out of business in 1721. Luckily, Sloane, who had recently purchased the land on which the Chelsea Physic Garden stood, was looking for a new gardener. Miller was hired and within just a few decades turned his patron's newly

Opposite An illustration of *Cassia bahamensis* from John Martyn's *Historia Plantarum Rariorum* showing a plant from the legume family described by Philip Miller as a new species in *The Gardeners Dictionary*.

Cassia Bahamensis, pinnis foliorum mucronatis, angustis, calyce floris non reflexo.
Gulielmo Sloane Armig.
Soc. Reg. Lond. Sodali.
Van Huysum pinx.
E. Kirkall sculp.

acquired garden into the envy of the world. Through an elaborate network of international correspondents, Miller ensured a constant flow of new plants into his workplace at Chelsea. The traveller Richard Pococke and William Sherard, the Levant Company's consul in Smyrna, sent seeds of eastern Mediterranean plants, whilst European gardeners from Florence, Leiden, Paris and St Petersburg supplied the latest botanical discoveries from their own countries and from distant places such as China, Canada and Ceylon.

TO

Sir *HANS SLOANE*, Bart.

PRESIDENT,

AND, TO THE

Council *and* Fellows

OF THE

Royal Society of *LONDON*,

FOR

Improving Natural Knowledge;

THIS

GARDENERS DICTIONARY

Is Humbly Dedicated by

Their moſt Obedient Servant,

Philip Miller.

Above An introductiory page from the first edition of *The Gardeners Dictionary*, published by Philip Miller in 1731 and dedicated to Sir Hans Sloane and the Royal Society.

Miller's greatest contribution to the development of Chelsea as a centre for horticultural innovation was the result of his connections in the British colonies in North America and the West Indies. The botanical riches of these parts of the world were relatively little explored but their potential value to British horticulture was enormous. With the help of his friend and avid gardener Peter Collinson, Miller procured a regular supply of plants from the eastern states of America collected by the Quaker naturalist and father of American botany, John Bartram. The plant collectors Mark Catesby and Thomas Dale gathered seeds of many ornamental trees and shrubs from Virginia, Carolina and the Bahamas. Scottish botanist William Houstoun dispatched numerous plants from Central America and the Caribbean, among them *Mimosa houstoniana* Mill., an elegant species of powder puff tree from Mexico named by Miller in honour of his esteemed colleague.

Many of the plants introduced by Miller at Chelsea were recorded in successive editions of his *Gardeners Dictionary*, where more than 1,000 were also described as new species. Their types, the dried and pressed specimens that remain scientifically important today for the naming of species, were preserved in Miller's own herbarium along with nearly 10,000 other plants. Sir Joseph Banks purchased this valuable collection in 1774, three years after Miller's death, and it later formed the nucleus of the General Herbarium at the Natural History Museum in London.

Between 1727 and 1739 Miller also presented over 1,400 specimens of various plants cultivated at Chelsea to his benefactor and the garden's landlord, Sir Hans Sloane. They are bound in 12 volumes of Sloane's herbarium and provide an insight into the most productive period of Miller's career, when plants from far-flung countries were arriving in Britain at an astonishing rate and the early editions of *The Gardeners Dictionary* were rolling off the press on an almost biennial basis. This carefully curated selection of only the rarest plants is an endearing expression of Miller's gratitude for his patron's support and a microcosm of botanical delightfulness. It is another rich source of scientifically important type specimens for the new plant species that Miller described. The majority of them are also found in Miller's own herbarium; but others such as the type of *Cassia bahamensis*, an attractive shrub from the legume family introduced by Mark Catesby in 1726, are not duplicated outside Sloane's collection. Miller identified them all with long and descriptive polynominal names that differ noticeably from their modern two-part equivalents because the Linnaean system of binominal nomenclature was not yet in use. Miller staunchly resisted this revolutionary idea of plant naming when it was first introduced by Linneaus in 1753; only eventually adopting it in the eighth edition of *The Gardeners Dictionary* published in 1768.

Many of the polynominals used by Miller for specimens in the Sloane Herbarium, for example *Linaria Hispanica procumbens foliis uncialibus flore flavescente pulchre striato labiis nigro-purpureis* Act. Phil. n. 412, correspond with the names published in papers in the *Philosophical Transactions of the Royal Society*. Those publications list specimens that were prepared from the plants cultivated at Chelsea during Miller's tenure and delivered to the Royal Society not by Miller himself but by the 'Demonstrators of Plants'. The specimens were made because a deed of conveyance, signed in 1722, transferred Chelsea Physic Garden to the Society of Apothecaries in perpetuity for an annual rent of £5 on condition that 50 specimens of plants grown there were to be presented to the Royal Society every year until the total of 2,000 was reached. In the end, no fewer than 3,750 of such specimens were collected, including the toadflax to

Left A type specimen of *Cassia bahamensis* (known today as *Senna ligustrina*) in the Sloane Herbarium, a Caribbean plant grown and collected by Philip Miller at Chelsea Physic Garden around 1727.

which Miller alludes in his polynominal. They were all transferred to the British Museum in 1781 and incorporated into the General Herbarium at the Natural History Museum in the late 1880s.

Some of Miller's specimens in Sloane's collection, like the *C. bahamensis* specimen, also include references to illustrations published in *Historia plantarum rariorum* (1728–1737), a magnificent work by the English botanist John Martyn, featuring 50 colour-printed plates of the most unusual plants grown at Chelsea and other grand gardens of England at the beginning of the eighteenth century. Many of the plates were later cited by Linnaeus in his *Species Plantarum* and plate number 41 serves today as the type of *Milleria quinqueflora* L., a tropical herb from the daisy family named in honour of Philip Miller. Other plants given by Miller to Sloane were illustrated in *The figures of the most beautiful, useful and uncommon plants* [...] (1755–1760), a two-volume book published by Miller as a supplement to *The Gardeners Dictionary*. Miller's brother-in-law, Georg Dionysius Ehret, one of the greatest botanical draughtsmen of all time, was responsible for many of the illustrations in this work.

Above An illustration of *Milleria quinqueflora* from John Martyn's *Historia Plantarum Rariorum* showing a plant discovered by William Houstoun in Central America and named by Carl Linnaeus in honour of Philip Miller.

It is somewhat ironic that for all his fame and glory we don't know what Philip Miller looked like. An elaborate engraving long thought to depict him and featured in many publications about him was recently proven to be a self-portrait of John Miller, a botanical illustrator and 'Demonstrator of Plants' working at Chelsea at the same time as Philip Miller. However, the lack of a portrait perhaps does not really matter; Philip Miller leaves as his legacy the spectacular plants he introduced to our gardens and it is through their beautiful blooms that he is best remembered.

GEORGE LONDON

MARK A. SPENCER

As the seventeenth century progressed, English people of wealth and high social status sought new ways to show their flair. As the arts and sciences flourished, the role of the garden took on new meaning. Gardens became a means of demonstrating good taste, wealth and an enquiring mind. To many, the key signature of a fine garden was good design. The country's aristocracy and the wealthier merchant class desired their own take on the great gardens of Versailles designed by André Le Nôtre, the famed Frenchman who also designed Greenwich Park for King Charles II. Home-grown garden design talent came in the form of George London (*c.* 1640-1714) and his apprentice, and later partner, Henry Wise. The London and Wise team became the most fashionable garden designers in England, their rather formal, largely symmetrical designs held sway for several decades until being swept away by the revolutionary schemes of Lancelot 'Capability' Brown. All of London and Wise's gardens were lost, although the garden of Hanbury Hall in Worcestershire has been restored. It is also probable that elements of London & Wise's design survive at Canons Ashby House in Northamptonshire.

London initially served as an apprentice to John Rose, gardener to King Charles II and the Earl of Essex. He then went on to work for Henry Compton, Bishop of London at his gardens in Fulham Palace. London also helped set up Brompton Park nursery and he later took control of the business on the demise of the other founding partners. The writer, gardener and diarist, John Evelyn described the nursery as 'the greatest work of the kind ever seen, or heard of, in either Books or Travels'. It was through this business that London was able to supply his many wealthy clients with the plants they desired. In ensuring his clients got the best, London needed to not only be an astute businessman and garden designer,

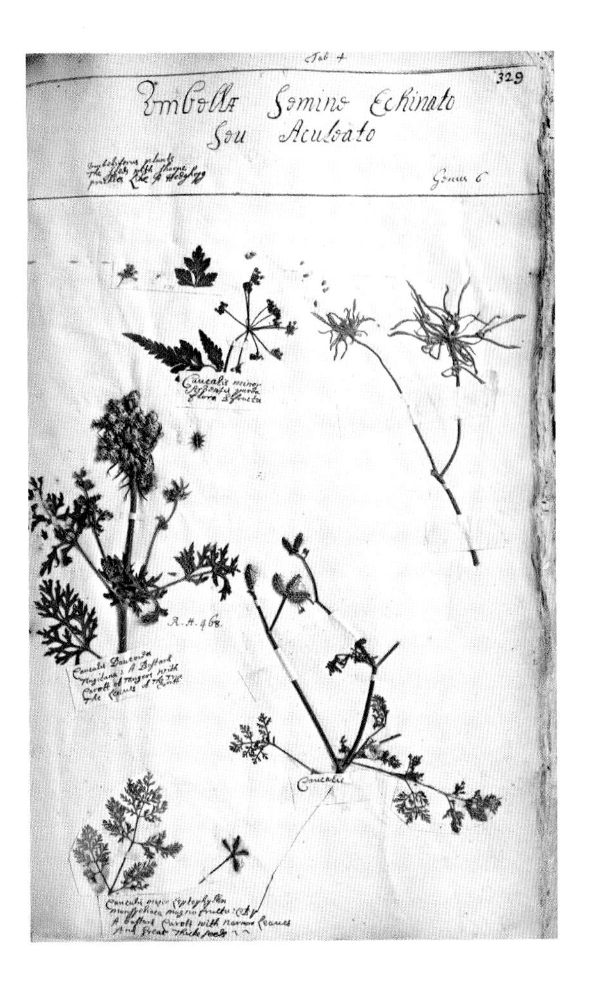

Above A specimen of small bur-parsley (*Caucalis platycarpos*) alongside other members of the carrot family (Apiaceae) collected by George London.

he also needed to know his plants. He was as adept botanically as he was horticulturally.

Historians have largely overlooked London's botanical knowledge, primarily because the surviving evidence for this is within the Sloane Herbarium and remains uncatalogued. There are two volumes of specimens that are of interest, although other volumes contain specimens that he gifted to members of the capital's botanical and gardening community. The first volume includes specimens that he presented to Leonard Plukenet. London's specimens in this voluminous book, the largest in the Sloane Herbarium, largely originated from his visit to Leiden's famous and influential *Hortus Botanicus,* where he met the eminent botanist Paul Hermann who held the Chair of Botany at the University of Leiden. The other volume appears to contain specimens that London collected throughout his life. Parts of this volume are arranged following the classification of the Oxford botanist Robert Morison whose works were soon to be eclipsed by those of that superlative natural historian, John Ray. On one folio, entitled '*Umbellae Semine Echinato seu Aculeato*' London references 'Tab 4' and 'Genus 6' of Morison's *Plantarum Umbelliferarum Distributio Nova* (1672), an important work that was the first to describe and classify all known members of the carrot family. London further describes the plants on the page as 'umbelliferus plants. The seeds with sharpe prickles like a Hedghogg'. There are several specimens on the page, one of which is identified by London as '*Caucalis maior leptophyllon monspeliaca magna fructu*: C.B.P. A bastard carrot with narrow leaves. And great thick seeds'. The 'C.B.P.' refers to Caspar Bauhin's *Prodromus theatri botanici* (1620) which, despite being many decades old, remained an important botanical reference work. The plant London had collected, pressed and annotated is small bur-parsley (*Caucalis platycarpos*), Now extinct as a wild plant in Britain, it appears to have been locally frequent in arable fields in the seventeenth century. It is not known who taught London his botany, but it is evident that he understood the literature and the plants. Indeed, he was so taken with Morison's methods that he copied Morison's unusual diagrammatic schema demonstrating the identification characters into his herbarium alongside the relevant plants.

Unsurprisingly for a nurseryman, London appears to have been interested in the seeds of plants. London presents specimens of medicks (*Medicago* spp.) and goes as far as to create a tabular arrangement of the fruits of these closely related plants to aid comparison. Botanists still depend upon examining their fruit when identifying medicks.

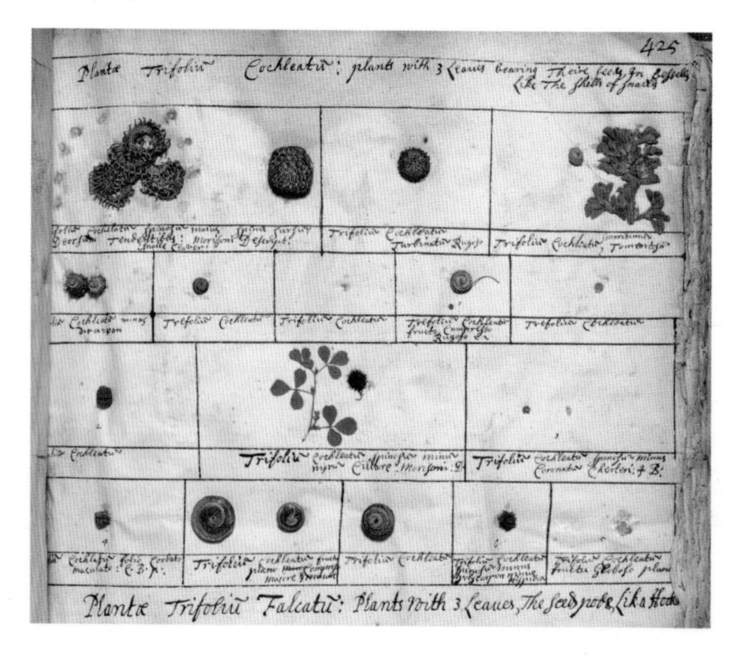

Left The fruit of various medicks (*Medicago* spp.), arranged by London to ease comparison.

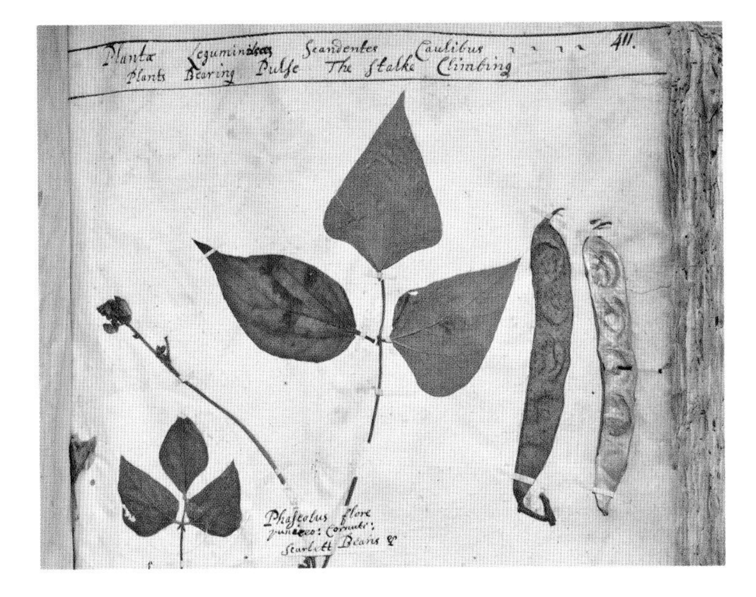

Left Now very familiar as a vegetable crop, the scarlet runner bean (*Phaseolus coccineus*) was once more prized as an ornamental.

Garden and vegetable plants that now seem prosaic are also present. The scarlet runner bean (*Phaseolus coccineus*) was introduced to Europe from the Americas shortly after Columbus's arrival there. In London's day, runner beans were still novelties that were largely grown for their exotic climbing habit and bright red flowers. DNA based research of well-preserved historic herbarium specimens such as these can provide important information on the origins and domestication of crops and other economically important plants. In turn, this information can be used to develop disease resistant and less environmentally impactful crops.

Prior to the mid-seventeenth century European gardens were largely stocked with plants originating from the continent or nearby regions such as the Middle East and North Africa. The expanding international trade (and ultimately colonial) networks pioneered largely by the Dutch and the Spanish resulted in many new plants arriving in Europe. Importantly, the activities of the Dutch East Indies Company (Vereenigde Oostindische Compagnie or VOC) allowed many plants from South Africa to be trialled in the botanic gardens at Leiden and Amsterdam. From these institutes, the ancestors of many important garden plants such as geraniums (*Pelargonium* spp.) and nasturtiums (*Tropaeolum majus*) spread to the early botanic gardens of London such as Bishop Compton's garden at Fulham Palace. One of the earliest to arrive in Europe was the night-scented geranium (*Pelargonium triste*) which has been grown in England since the 1630s. It was known to London and other gardeners as 'Sweet Indian Cranes Bill', reflecting the widespread confusion about the origins of these South African plants. This confusion was because the plants originally arrived via the VOC and it was therefore assumed that they had originated from the East Indies.

Above The night-scented geranium (*Pelargonium triste*) was the first *Pelargonium* to be introduced from South Africa. Several of its wild relatives are the parents of these popular garden plants.

Increasing contact with the Americas saw the introduction of exciting novelties like passionflower (*Passiflora* spp.). These plants were no doubt popular with George London's clients not only for their beautiful and complex flowers but for the symbology relating to the crucifixion of Jesus. The cultivation of these plants would have required winter warmth and they would have been grown in the relatively primitive stovehouses of the era. Alongside passionflower, other American exotica such as Indian shot (*Canna indica*) were grown. A large proportion of the plants mounted in London's *hortus siccus* are fragmentary, as are many others in the Sloane Herbarium. Generally, they would not qualify as good quality herbarium specimens today. We now attempt to collect enough material to reflect all the relevant characters that aid identification and pass on a sense of the plant's form. This was not necessarily the objective in the late seventeenth and early eighteenth centuries; specimens often appear to have been collected as *aide-mémoires*, something to remind the collector or viewer of what the plant's name is. In many cases, the plants were also so rare that it was not possible or desirable to sacrifice a significant part of the plant.

Above A decorative arrangement of flowers and leaves from George London's herbarium.

London's personal herbarium is one of the more complex in the Sloane Herbarium and it is made up of several distinct parts. Despite this complexity, this overlooked collection provides a unique and important insight into late seventeenth century English horticulture's experimentation with novel plants arriving on these shores. Some of the herbarium appears to be remnants of previous volumes, especially those following the works of Robert Morison. One of the more anachronistic sections contains several pages of purely decorative work. These pages are made up of ornamental flowers, many of them double, and leaves. There are no scientific names or annotations and the identity of the creator is unknown. Perhaps it was George London himself, but my personal belief is that they were created by a child, purely with the aim of providing pleasure. London clearly desired that his herbarium be viewed by others, at the bottom of one page, he wrote 'Accept my endeavours, Pardon my failings and farewell.'

COLLECTORS IN THE FIELD: BRITAIN AND EUROPE

MARK CARINE AND ROBERT HUXLEY

When Sloane acquired William Courten's herbarium in 1702, he catalogued it meticulously. It was a collection rich in material from the British Isles and continental Europe and its specimens were carefully cross-referenced to John Ray's *Historia Plantarum*. Sloane's cataloguing revealed no striking new discoveries among those British and European plants; in most cases, Courten's specimens were readily matched to those already described in the *Historia Plantarum*. Despite this, knowledge of the flora of Britain and Europe at that time was far from complete. There were still many species to discover in the seed plant flora and in groups such as the bryophytes and the lichens, knowledge of the flora was even more limited.

By 1680, around the time when Sloane first started to assemble his collection, just under 1,000 species, or around 60% of the native seed plants of Britain and Ireland had been recorded. By the time of Sloane's death in 1753, more than 200 further species had been added, an average of around three each year. The Sloane Herbarium includes a significant number of those new records. It includes, for example, the earliest record of *Gentianella germanica,* the striking Chiltern gentian, a species that in Britain is largely restricted to the chalk hills north and west of London and that was first collected in 1707 by Richard Tilden. It also includes new discoveries for Britain made by Sloane himself. Sloane was the first to record the glasswort, *Sarcocornia perennis,* in England and the first to record the water-starwort, *Callitriche platycarpa.* The label that accompanies a specimen of the latter in James Petiver's herbarium explains that 'The first discovery of this Plant to be a native of England we owe

Opposite Seaweed specimens from Adam Buddle's collection.

to the ingenious Physician and Botanist Dr Hans Sloane who observed it in a bog on Putney Heath, June 4, 1691'.

For many continental European countries, knowledge of the flora among Sloane and his contemporaries was more rudimentary. When Sloane acquired specimens from Spain and the Balearic Islands, for example, many were new to his collection and many did not feature in Ray's great *Historia Plantarum*. The Mediterranean region, one of the world's biodiversity hotspots, was very much a botanical frontier for Sloane.

The map of collecting localities on p.13 reveals, perhaps not surprisingly, that the region encompassing the British Isles and continental Europe was more densely sampled in the Sloane Herbarium than anywhere else. There are more sampling locations recorded and there are more contributors of specimens than from other parts of the world. Many of those individuals who collected in Britain and Europe collected only there and nowhere else. James Sutherland, the superintendent of the 'Physick Garden' at Edinburgh, contributed 'several uncommon Northern Plants, which he had gathered wild in Scotland'. Richard Wheeler, a correspondent of Petiver, contributed plants from Norway. Petiver noted '...my kind and hearty Friend, hath for many Years past, and doth still continue to send me all the Plants ... which he can get or procure me, about Long-Sound in Norway: many ... being rare with us, and only to be met in the North of England'. Johann Jakob Scheuchzer, a professor of mathematics and physics in Zürich, whose son would go on to work in Sloane's library, sent plants from the Alps. Gedeon Bonnivert, a soldier born in France but who served in the English Army,

Below The Chiltern gentian (*Gentianella germanica*). The earliest record of this species for Britain and Ireland is a specimen in the Sloane Herbarium.

travelled more widely, collecting where he was stationed and contributing specimens from England, Ireland, Belgium and the Netherlands.

For others, Britain and Europe was where they first cut their collecting teeth before moving on to other, more remote parts of the world. The surgeon Sylvanus Landon first sent plants to Petiver from Spain before travelling to Asia and contributing the only material in the Sloane Herbarium collected on the island of Pulau Flores in Indonesia. Sloane himself first collected in Britain and France before going on to make his most significant field collections during his voyage to Jamaica.

In this chapter we look at three individuals who made significant contributions both to the Sloane Herbarium and to the discovery of the flora of Britain and Ireland, namely the Suffolk clergyman Adam Buddle, the Yorkshire physician Richard Richardson and the Welshman Edward Lhywd, the second Keeper of the Ashmolean Museum in Oxford. We also feature two of the numerous individuals who contributed specimens from further afield in Europe, and specifically from the Mediterranean: Bruno Tozzi, a Benedictine monk in Tuscany, Italy; and Salvador i Riera, an apothecary from Barcelona, Spain. Both were part of James Petiver's extensive network of correspondents and their collections came to Sloane after Petiver's collection was acquired in 1710. Both men were sources of new species for the Sloane Herbarium.

Above Gedeon Bonnivert was a soldier who contributed specimens to the Sloane Herbarium from England, Ireland, Belgium and the Netherlands. This specimen of a green hellebore (*Helleborus viridis*) was collected in Flanders, Belgium, where he was serving as a Lieutenant in the English army.

REVEREND ADAM BUDDLE

MARK A. SPENCER

Every summer our towns and cities, and increasingly our countryside, are festooned with the purple flowers of a shrub that celebrates a man whose life and works are now largely forgotten. The plant is an invasive non-native shrub from China known as buddleia or butterfly bush. The scientific name for this plant that reached our shores in the late nineteenth century is *Buddleja davidii*. This purple peril is named after Reverend Adam Buddle (1662-1715). Today he is an almost forgotten man and all that remains is *Buddleja*, a name first used by the botanist William Houstoun and taken up by Carl Linnaeus to commemorate him. Buddle was not only a churchman, he was a botanist who compiled a comprehensive herbarium and wrote an unpublished manuscript on the wild plant life of Great Britain and Ireland. He was viewed as a man of considerable expertise by his fellow botanists across Europe.

What little is known about the man himself has been garnered from county and parish records as well as the collected correspondence compiled by Sir Hans Sloane that are now held at the British Library alongside Buddle's unpublished manuscript, the *Methodus nova stirpium Britannicarum ex methodis et Raij et Tournefortij longe optimis collatis correcta cum nominibus et synonymis autorum maxime celebrium additis*. The *Methodus* manuscript acts as an index and catalogue for the specimens in Buddle's herbarium. The work was aimed at being an improvement on the taxonomic systems of the English natural historian John Ray and the French botanist Joseph Pitton de Tournefort, who were luminaries of the natural sciences at the time.

Buddle was born in Deeping St James in Lincolnshire in 1662, two years after Sloane. Showing a scholarly flair, in 1678 he entered Catherine's Hall (later St Catherine's College), University of Cambridge. He became a fellow of the college in 1686 but his

Left Nearly 40 years after Buddle's death, Carl Linnaeus named this plant *Buddleja americana*, familiar to Sir Hans Sloane from the Rio Cobre in Jamaica, in honour of Rev. Adam Buddle.

refusal to sign the Oath of Allegiance to King William resulted in his ejection in 1691. It is likely that he started compiling the herbarium shortly afterwards. In 1695, at Friston in Suffolk, he married Elizabeth Eveare. The couple moved to Henley in Suffolk, where in 1695 their daughter Elizabeth was born, followed in 1697 by Adam, who died after a few days. In 1699 another son, also Adam, was born in St Giles, Cripplegate, London, of whom there is no further record and it is possible he too died in early childhood. It is probable that Buddle received an inheritance from an uncle, Adam Buddle of Hadleigh, Suffolk. He was ordained rather late in life, at the age of 40, in Ely Cathedral and obtained a living at North Fambridge in Essex in 1703. Later that year he was appointed Chapel Reader at Gray's Inn, London, where he retained a chamber until his death in 1715. He was buried at St Andrew's, Holborn, which is also the resting place of the herbalist John Gerard and the botanist Christopher Merrett.

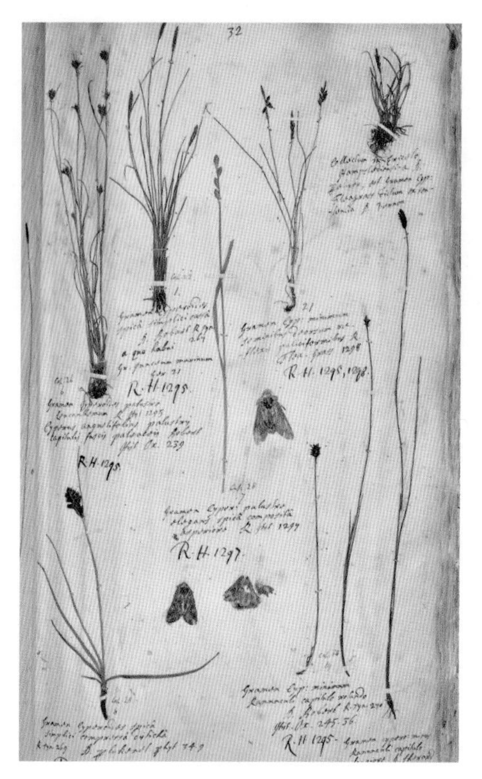

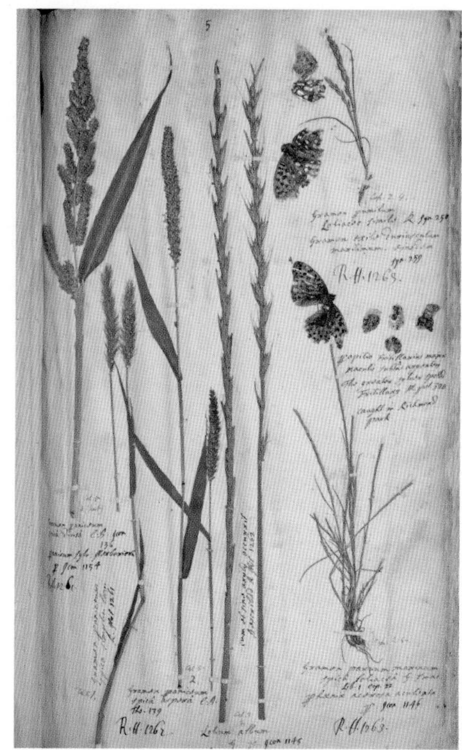

Above Pages from Buddle's herbarium, including flea sedge (*Carex pulicaris*) collected by Buddle and James Petiver from Hampstead Heath and dark green fritillary butterflies (*Argynnis aglaja*) collected by Buddle from Richmond Park.

In his will Buddle stipulated that his herbarium and manuscript be given to Sloane, who by that time had already acquired an extensive herbarium compiled of his own collections as well as those of others, notably Leonard Plukenet. It took some time for Sloane to receive the herbarium, for at the time of Buddle's death it was in the possession of James Petiver. Shortly before his death, Buddle's wife Elizabeth anxiously wrote to Petiver, pleading: 'I desire you will send by this bearer Mr Buddle's volume of plants and his manuscripts for he has charged me to send them to Dr Sloane which I desire to do this day for to my great sorrow I find I must lay aside all hopes of his recovery.' On receipt of the herbarium, Sloane first lent it to Jacob Dillenius, who made great use of it in the final edition of Ray's *Synopsis Plantarum Britannicarum* (1724) and his own masterpiece *Historia Muscorum* (1741). Sloane then passed it back to James Petiver.

The herbarium is full of treasures. Many remain overlooked because it has not been fully catalogued since Buddle completed the *Methodus* in 1707. It is believed that Buddle started compiling the herbarium in 1692 and that he commenced his work on the grasses and their relatives. That he started the herbarium by working on grasses, a group that botanical beginners are notoriously fearful of, suggests he was already a very competent botanist. The herbarium originally consisted of 13 volumes although these were rebound, in sequence, into four after they came into Sloane's possession. Sloane's *Hortus Siccus*, HS 125 (Buddle's volume XII) contains Buddle's fine collection of grasses, sedges and rushes. One of these, a specimen of flea sedge (*Carex pulicaris*) was collected on Hampstead Heath near London, in the company of James Petiver ('in Ericeto Hampstediensi a D. Petiver'). At the time, Hampstead Heath was still a heathland dominated by heathers and gorse and had been a favoured botanizing location since the time of Matthias de l'Obel in the mid-sixteenth century. This volume is also notable in that it contains many pressed insects, some of which are cited by Petiver in his *Musaeum* and *Gazophylacium*. Among them are two dark green fritillary butterflies (*Argynnis aglaja*) accompanied by text in Buddle's hand copied from page 320 of *Musaeum Petiverianum* with an additional note explaining that they were 'caught in Richmond Park'.

Curiously, HS 125 is the only volume of Buddle's containing pressed insects,

Below Although it looks like seaweed, hornwrack (*Flustra foliacea*), top left, is an animal. These specimens share the page with various seaweeds including the brown algae (*Cystoseira tamarisciifolia*) collected by a Mr Stevens from Menheniot in Cornwall.

although there are insects in other volumes of Sloane's great herbarium. The remaining volumes associated with Buddle contain algae (especially seaweeds), fungi and plants. The algal and fungal collection (H.S 114; Buddle's volume I) contains surprises for the modern botanist, since many of the specimens are not algae or fungi. Some pages contain colonial animals known as hydrozoans and bryozoans. In the wild, their apparently immobile state meant that for centuries they were 'lumped' with algae. Even today the common bryozoan *Flustra foliacea* is frequently mistaken for seaweed by beachcombers. Two of Buddle's *Flustra* specimens share a page with seaweeds, including a specimen of the brown algae *Cystoseira tamarisciifolia*, the bushy rainbow wrack, sent 'From Mr Stephens [Stevens]' to Buddle from 'out of Cornwall'. These specimens, like most of the other specimens within Buddle's herbarium, are also annotated with bibliographic references that link them to information provided in Ray's *Historia Plantarum* or other key taxonomic works of the era.

Buddle's herbarium is full of specimens sent to him by English and Welsh botanists. In addition to the previously mentioned Reverend Lewis Stevens of Menheniot, Cornwall, Buddle received specimens from across Great Britain and from Sligo in Ireland. Particularly significant collections made their way to Buddle from Snowdonia, then a very remote location, via the naturalist Richard Richardson. The most celebrated of Buddle's acquisitions from Richardson is the Snowdon lily (*Gagea serotina*, formerly *Lloydia serotina*). It is one of two specimens collected by Richardson in the presence of the discoverer, Edward Lhwyd; the other is in the herbarium of James Petiver (see essay on Lhwyd).

There was one group of plants that Buddle particularly excelled at: mosses, liverworts and their relatives. His taxonomic expertise in this regard was such that he was affectionately known as 'the top of all the moss-croppers' by his friend William Vernon. Sloane obviously considered Buddle's skills to be worthy, as he appears to have sent Buddle specimens from his time in Jamaica. Buddle also received a *Selaginella* specimen collected by Archibald Stewart

Opposite This Snowdon lily (*Gagea serotina*) was sent to Buddle by the prolific naturalist Richard Richardson who collected the plant with Edward Llwyd.

from the Kingdom of Scotland's ill-fated Darien Scheme colony known as 'Caledonia' in Panama. The *Selaginella*, and a duplicate in Petiver's herbarium (see p.30), are one of a handful of specimens from 'Caledonia' that have survived to the present day. .

Buddle's herbarium and manuscript are fine works and it should be lamented that he remains largely forgotten. James Edgar Dandy commented in *The Sloane Herbarium* (1958) that, 'Among the botanists of Sir Hans Sloane's time the Rev Adam Buddle stands out for his sincere interest in the taxonomy of British plants. His single-minded devotion to British botany, which led him not only to compile a wonderful herbarium arranged in systematic order, but to complete a manuscript ready for publication, did not meet with its due reward.' Thankfully, Buddle's reputation held fast for a few decades and in *Species Plantarum* (1753) Linnaeus renamed a plant known to Sloane in Jamaica as '*Verbasci folio minore arbor, floribus spicatus luteis, seminibus singulis oblongis in singulis vasculis siccis*' as *Buddleja americana*.

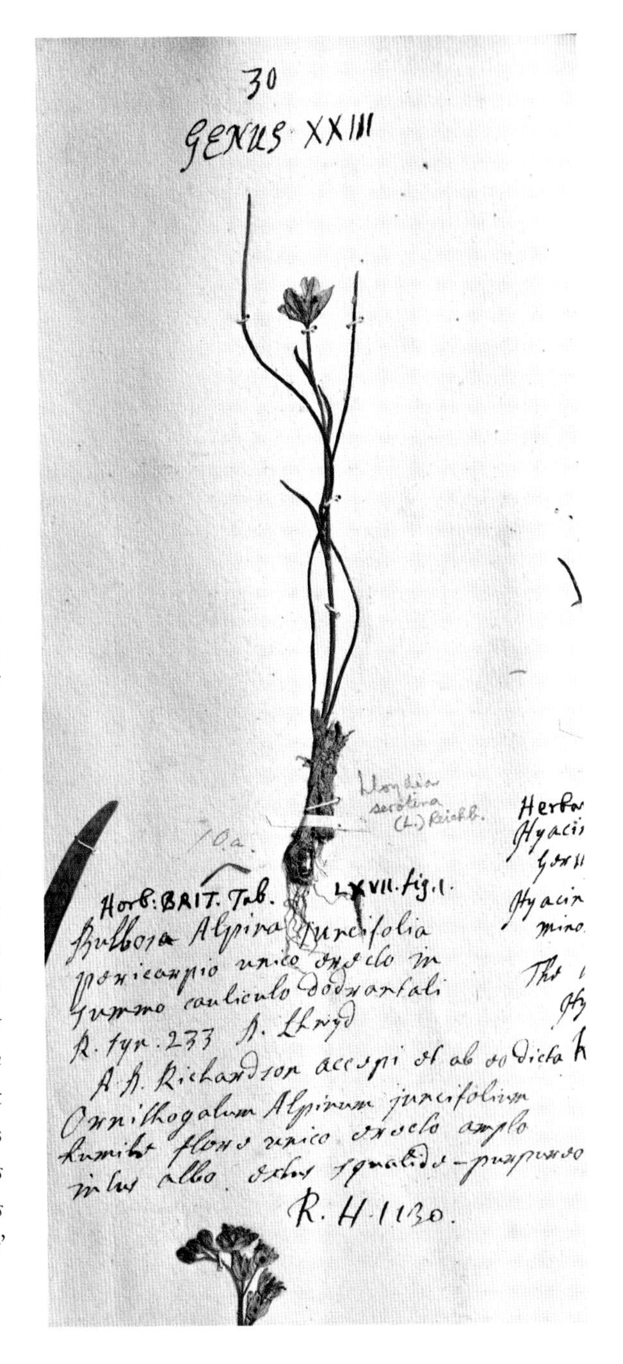

EDWARD LHWYD (LHUYD)

FRED RUMSEY

Edward Lhwyd was born in Loppington, Shropshire, in September 1660, the illegitimate son of Edward Lloyd of Llanforda, near Oswestry, and Bridget Pryse of Llansantfraid, Cardiganshire, who had met while Lloyd, an ardent Royalist, was in hiding during The Protectorate. A marriage was apparently arranged but never took place, although Lhwyd was to maintain a close connection to his maternal relatives throughout his life. It was from his father, described as an 'able and versatile man, [although] eccentric and dissolute' that Lhwyd inherited his interest in plants and the natural world. He seems to have received some botanical instruction in his teens from Edward Morgan, a gardener once employed at the Westminster Physick Garden and who had been with Thomas Johnson in August 1639 when the botanical riches of Snowdon were first revealed. Lhywd would later go on to make a significant contribution to knowledge of the flora of that region.

Lhwyd matriculated at Jesus College, University of Oxford, and remained there for five years, supplementing his meagre living by becoming an assistant at the newly founded Ashmolean Museum in 1684 and 'Register of the Chymicall courses of ye Laboratory', both of which will have seen him under the wing of the Professor of Chemistry and first Keeper of the Museum, Dr Robert Plot. Plot's work on 'formed stones', or fossils, undoubtedly kindled Lhwyd's great interest in this area and it is the study of fossils to which he would go on to devote most of his time. Lhwyd is credited with being the founder of British palaeontology, his fossil collections being the earliest surviving. He was the first to collect a trilobite, creatures he thought 'must doubtless be referred to the skeleton of some Flat-Fish'. He was also responsible for the first scientific description and naming of the most iconic of fossil organisms, the dinosaur, or at least the tooth of one that we would

Above The cliff of Clogwynn du'r arddu on Snowdon, where Lhwyd found many of its botanical rarities and still the best place to see the Snowdon lily, originally named *Lloydia serotina* for him.

now recognize as a sauropod. Although Lhwyd failed to graduate, his later work would see him honoured with an MA *honoris causa* by the University in 1701 and he would go on to become a Fellow of the Royal Society in 1708. He died of pleurisy at the Ashmolean on 30 June 1709.

In the late spring and summer of 1682, shortly before entering Oxford, Lhwyd made an extensive botanical tour of North Wales, taking in the mountains of Cader Idris and Snowdon, reckoning in a letter to his kinsman David Lloyd to have found six species on the latter that 'have not been observ'd by Mr. Ray to be Natives of England'. His list of plants seen on Snowdon on 24 August 1682 contains many species he would go on to list in his account of 'more rare plants growing in Wales', published in Edmund Gibson's revised *Camden's Britannia* in 1695.

Lhwyd left a further list, *A Catalogue of some plants observed at Creigiau'r Eryran*, dated 24 August 1688, at 'Lhan Berrys' [Llanberis] 'for the use of such as came simpling'. This list came to the attention of John Ray who recognized that its creator was 'no trivial herbalist, but a man of good skill in plants'. Plot subsequently wrote to Lhwyd

Left A specimen of alpine saxifrage (*Micranthes nivalis*) in Petiver's herbarium, its discovery credited to Lhwyd and labelled by Petiver 'Mr Lhwyd's Welsh-Pride'. The same folio has the first recorded *Saxifraga spathularis* and *S.* × *polita* from Ireland.

Opposite top A letter from Richard Richardson preserved in Petiver's *Hortus Siccus Anglicanus*, Vol. III describing his discovery, with Lhwyd, of the Snowdon lily (*Gagea serrotina*) in flower.

Opposite bottom Richardson's specimen of the Snowdon lily (bottom left) collected with Lhwyd in early June 1700, the first time it had been seen in flower.

enclosing a letter from Ray 'who tells me in these very words, *that what you shall communicate will be so acceptable to him, that he shall look upon it as the greatest ornament of his book*'. Lhwyd responded by sending Ray a package of plants, many species of which were described for the first time in Ray's *Synopsis Methodica Stirpium Britannicarum*, the first true British flora, published in 1690. One of Lhwyd's plants documented in that work, *Isoetes lacustris*, was the first record of this ancient group of plants and it featured as one of only two illustrations in that work, based on a sketch by Lhwyd. For many of Lhwyd's other discoveries it wasn't until the third edition of Ray's *Synopsis Methodica*, published in 1724, that they would be figured, years after both men's deaths. One such plant was *Micranthes nivalis*, first discovered on Snowdon in 1639 but made familiar by Lhwyd's collections and rather pleasingly described by Petiver as 'Mr. Lhwyd's Welsh-Pride'.

In Autumn 1689 Lhwyd wrote to the naturalist and physician Martin Lister, 'I have some patterns of our choisest Snowdon plants yt I can spare: & if you please to accept of ym either for your own use, or to oblige any friend, I shall send ym to you: perhaps Dr

Sloane or Dr Plucknet may have some choice plants to spare in exchange.' Lhwyd's collections in the Sloane Herbarium today are scattered through a dozen or more volumes, including the collections of Plukenet (a near neighbour of Lister), of Courten (Plukenet's contemporary at Westminster School) and those of Buddle. They are particularly strong on his Snowdon finds, which include the majority of the new and rare plants given in his Llanberis list.

Lhwyd's powers of observation were acute, as evidenced by the nature and quality of his botanical finds. To demonstrate this, one does not need to look further than the beautiful little plant named by Salisbury to honour him, *Lloydia serotina* (which, cruelly, modern molecular studies have seen re-assigned to the genus *Gagea*). As anyone who has spent time hunting for this great rarity on some of the more basic volcanic rock faces high in the Snowdon range could tell you, finding this plant even when you know where to look is difficult. Its small chive-like leaves, sufficiently similar to the grasses with which it occurs, make its detection a challenge; this is exacerbated by its restriction (now at least) to more inaccessible rocks where close observation is not possible. Its flowers are short-lived and they are sparingly produced. That Lhwyd found this plant out of flower speaks volumes. From a note preserved in HS 152, Richard Richardson tells us of when he found the plant in the company of Lhwyd in the beginning of June 1700, 'being the first time he saw it in flower'. It is uncertain when Lhwyd had first discovered it but it was recorded by him in

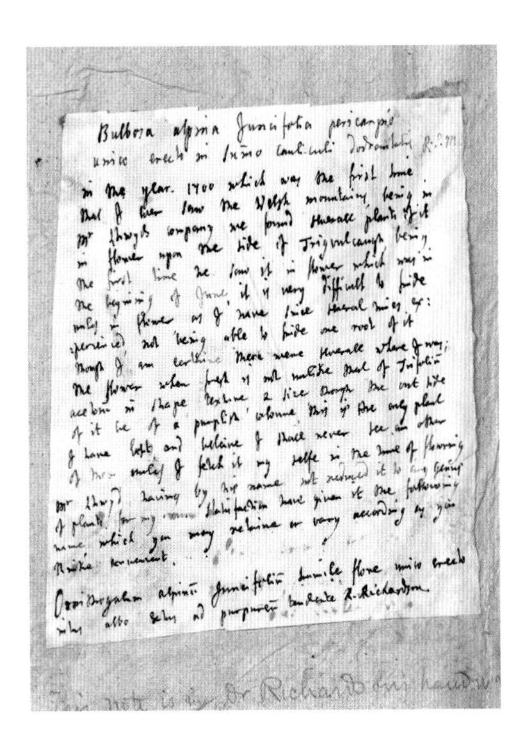
Bulbosa alpina Juncifolia pericarpio
unico erecto in summo cauliculi [illegible]
in the year. 1700 which was the first time
that I ever saw the Welsh mountains being in
Mr Lhwyds company we found severall plants of it
in flower upon the side of Trigvylchau being
the first time he saw it in flower which was in
the beginning of June, it is very difficult to finde
unless in flower as I have since severall times experienced not being able to finde one root of it
though I am certaine there were severall where I was;
the flower when fresh is not unlike that of Trifolium
acetosum in shape texture & size though the out side
of it be of a purplish colour this is the only plant
I have left and believe I shall never see an other
of them unless I fetch it my selfe in the time of flowering
Mr Lhwyd having by his name not reduced it to any genus
of plants for my [illegible] satisfaction have given it the following
name which you may retaine or vary according to your
[illegible] convenient.

Ornithogalum alpinum Juncifolium [illegible] flore unico erecto
[illegible] albo [illegible] ad purpuram tendente R. Richardson

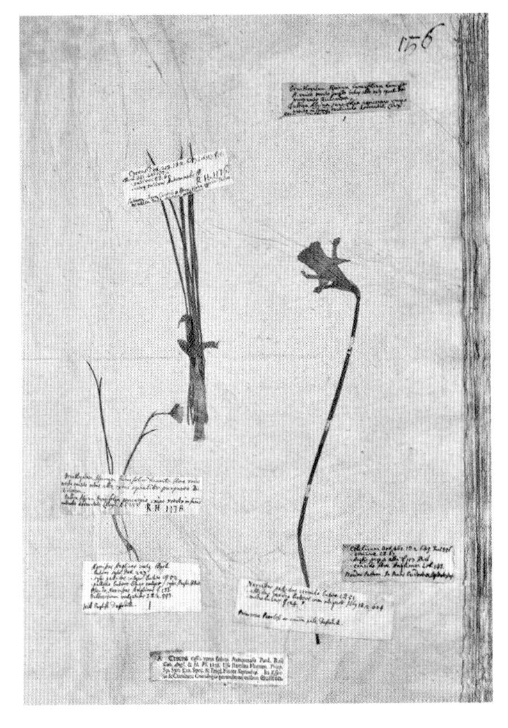

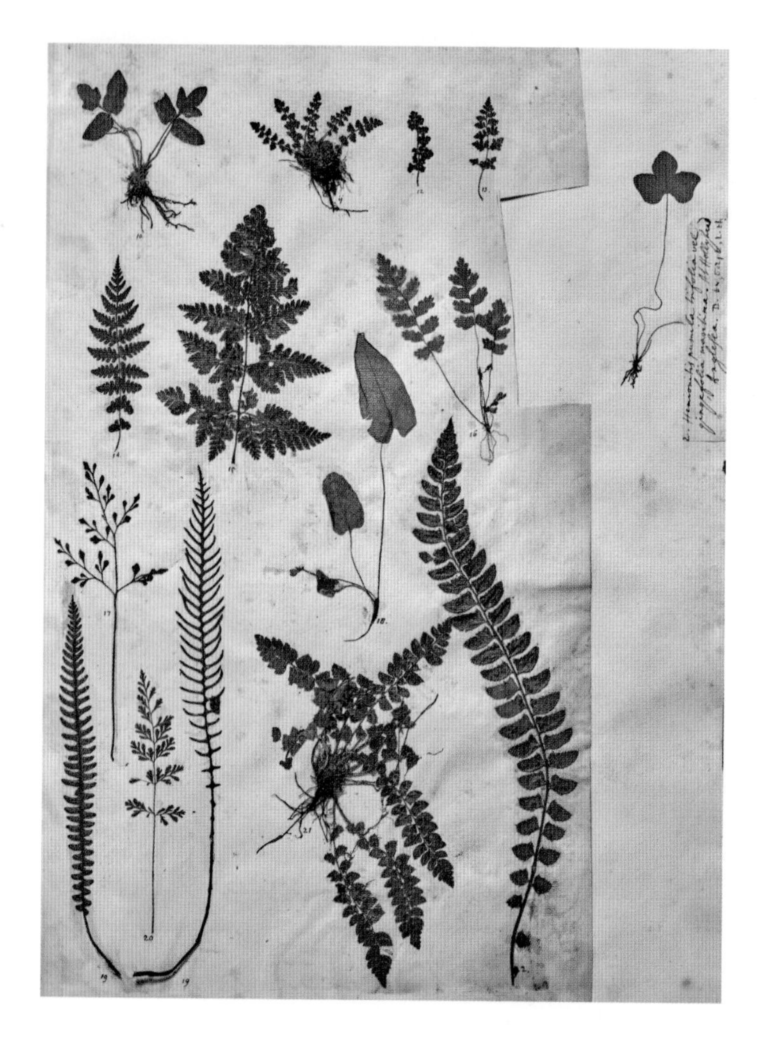

Left A selection of Welsh ferns in Courten's herbarium, collected by Lhwyd in the 1680s, several representing the earliest British records. Most are still extant precisely where Lhwyd found them.

1695 in his *Camden's Britannia* catalogue and a letter sent by Jacob Bobart to Lhwyd on the 23 September 1688 suggests an earlier awareness: upon the receipt from Lhwyd of plants from Snowdonia, Bobart laments 'that I am not soe fortunate as to find that bulb which in your letter you direct me to at the top of the biggest box, wch I search'd and research'd grass by grass and can by noe means find'. The missing plant was almost certainly the *Gagea*, the only bulbous alpine in the British flora. It is not impossible that Lhwyd may even have been aware of the species as early as 1682 when he wrote of 'several other plants, wch because they were not then in flower, I knew not whither to reduce'.

Lhwyd visited Ireland in 1700, drawn to the mountains of Kerry, Connemara and Sligo. He was to see a significant number of those Irish plants that would excite and exercise biogeographers centuries later, plants found only here and then again in Iberia – the Hiberno-lusitanian floristic element. His specimens of plants, such as St Patrick's cabbage (*Saxifraga spathulata*) are the earliest known Irish plant collections. They include the first records of three species, including St Daboec's Heath (*Daboecia cantabrica*) and the extremely rare endemic fringed sandwort (*Arenaria ciliata* subsp. *hibernica*). He collected the latter 'on the mountains of Ben Bulben and Ben Buishgen', still its only known location.

In addition to his palaeontological and botanical achievements, Lhwyd was a pioneering linguist. His own fervent desire to see the preservation of his native Welsh language was in part responsible for his chosen spelling of his own name, picked also to distance himself from the 'Lloyd' of his father and half-brothers; its various forms and transliterations, as evidenced in the Sloane collections, giving archivists and indexers many problems.

In his biography of Lhwyd, Robert T. Gunther summed him up: 'Not gifted with an impressive personality, and with so slender an endowment of worldly goods that he must often have gone hungry, he won his successes by steady application and enthusiasm, and by maintaining correspondence with the most learned contemporary authorities...' Sloane reportedly called him 'ye best naturalist now in Europe'. Few men of such great scientific achievements have been more neglected. Gunther seems unaware of the survival of Lhwyd's botanical specimens, scattered as they are within the Sloane Herbarium but his 'Patterns', as Lhwyd referred to specimens, continue to vouch for his important role in the discovery of the British and Irish floras and serve as a testament to this remarkable man.

RICHARD RICHARDSON

VICTORIA PICKERING

Richard Richardson was a physician and a great collector of plants, fossils and curiosities. He also created a large library that spanned his interest in antiquarianism, botany, geology, medicine and natural history. Like others in his botanic circle, Richardson had been educated at Oxford (University College) and he would eventually settle where he had been born in North Bierley near Bradford, practising as a physician. Richardson was captivated by the plant world. He travelled across various parts of Britain searching for rare specimens, and exchanged plant material with well-known and lesser known botanists around the world. Richardson had also studied abroad, spending time at the University of Leiden, where he lived with professor of botany Paul Hermann. His own garden at Bierley Hall contained an array of native and exotic plants.

Over a period of 40 years, Richardson and Sir Hans Sloane exchanged numerous lengthy and often happy correspondences, reflecting something of the close friendship that developed between these two physicians. One such extract from a letter written by Sloane to Richardson includes Sloane telling Richardson that:

> *I am very sorry that yor stay in London both times I have seen you here have been so short, that I have not had that opportunity of conversation with you that I could have wish'd for and particularly yor thoughts of many fossils, birds, eggs, plants & other naturall rarities & antiquities which I have gathered*

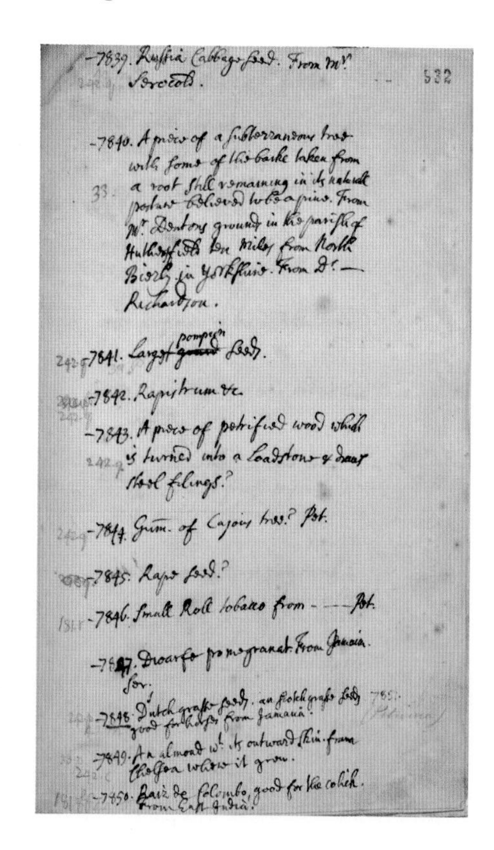

532

-7839. Russia Cabbage seed. From Mr. [illegible].

-7840. A piece of a subterraneous tree with some of the barke taken from a root still remaining in its natural posture believed to be a pine. From Mr. Denton's ground in the parish of [illegible] [illegible] miles from North Bierly in Yorkshire. From Dr. Richardson.

-7841. Largest ~~[illegible]~~ pompion seed.

-7842. Rapistrum &c.

-7843. A piece of petrified wood which is turned into a Loadstone & draws steel filings?

-7844. Gum. of Cajous tree? Pet.

-7845. Rape seed?

-7846. Small Roll tobacco from ---- Pet.

-7847. Dwarfe pomegranat. From Jamaica.

-7848. Dutch grass seeds. an English grass seeds good for horses from Jamaica.

-7849. An almond wt its outward skin from Chelsea where it grew.

-7850. Raiz de Colombo, good for the colick. From East India.

Below A page from one of Sloane's catalogues listing the objects in his expansive Vegetable Substances collection. It shows descriptions of botanical material sent from Richardson and the numbers listed correlate these entries with numbered boxes containing the specimens.

together. However I live in hope that some time or other I may be so happy. In the mean time to supply that letters are the best remedies and I should be extreamly glad now and then to hear from you and will promise to return you any accounts I think may be for yor entertainment.

Sloane clearly valued Richardson's thoughts and opinions on matters of natural history, and the tone of their exchanges reveals a friendship. Richardson and Sloane exchanged botanical news on new publications, people and their discoveries, as well as news on their own health and detailed accounts of various medical cases (relating to their patients) on which they both offered opinions.

Richardson and Sloane's exchanges were not confined to the written word. They also sent each other natural history specimens and books. Richardson was more than happy to 'hunt for Natural Curiositys for dr Sloane' and Sloane was likewise 'extreamly obliged' to him for his various samples and specimens, or 'letter[s] & present[s]', 'which gave [Sloane] a great deal of pleasure in turning over'. While they were both alive, Richardson contributed plant specimens to Sloane that were placed in his herbarium. He sent plants from Yorkshire and there is also 'An Herbarium vivum gathered by Dr Richardson in Holland out of the gardens, given to S.H.S. by Mr Vernon'. These specimens were gathered during the three years that Richardson spent

Below Shown here is Richardson's specimen of the Killarney fern (*Trichomanes speciosum*). It was the first record for this species and was likely the specimen that formed the basis for the figure in Ray's *Synopsis Methodica Stirpium Britannicarum* (1724). The fern can be found in the same locality today.

living in Leiden with Hermann. Originally mounted on small sheets, these specimens would be mounted onto larger sheets so that they could be incorporated into Sloane's herbarium.

Richardson also contributed material to Sloane's collection of Vegetable Substances. This is a collection comprised of over 12,500 botanical specimens that have been sealed into small ornate glass-and-wood boxes. The 13 samples provided by Richardson were mostly plant fossils and one example included 'A piece of subterraneous tree with some of the barke taken from a root still remaining in its naturall posture believed to be a pine'. It had supposedly come 'From Mr. Dentons grounds in the parish of Hutherfield ten miles from North Bierly, in Yorkshire'.

Sloane returned these favours. He sent samples to Richardson such as a seedling of the cedar of Lebanon, which grew to considerable splendour at Bierley Hall. Sloane also sent copies of the Royal Society's *Philosophical Transactions* and other natural history books. Contained within the letters he wrote to Richardson were Sloane's professional opinions on matters of medicine and treatments with regards to Richardson's patients. These exchanges, along with the botanical specimens collected and cultivated by Richardson, and which now form part of the Sloane Herbarium, reveal something very important about how Sloane was able to create such an impressive and expansive botanical collection. It was through his friendships with individuals such as Richardson that Sloane was able to benefit from a wide network of correspondents and therefore be involved with a wide-ranging exchange of British and European natural material. Sloane's relationship with Richardson reveals that friendship could influence the way a botanical collection was formed.

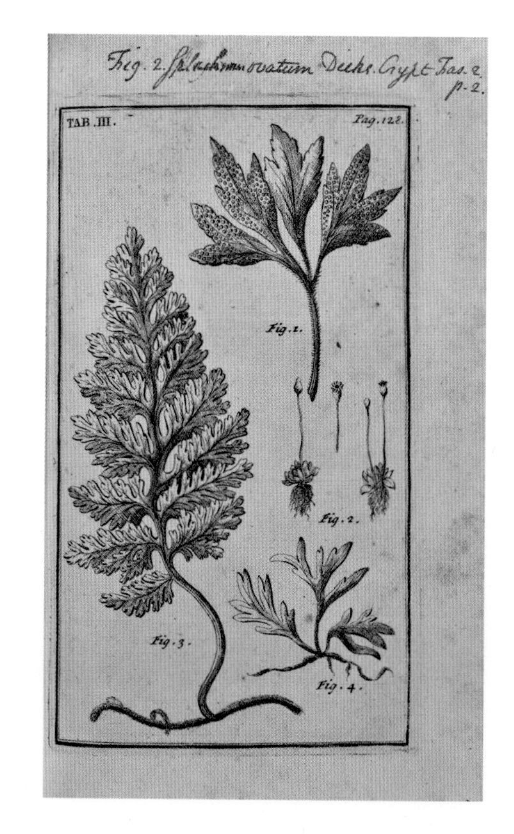

Below This illustration of the Killarney fern (*Trichomanes speciosum*), found in Ray's *Synopsis Methodica Stirpium Britannicarum* (1724) was based on the specimen collected by Richardson in Yorkshire (see previous page).

Above During his time at the botanical garden in Leiden, Richardson collected a variety of plants including this cultivated specimen of a flame lily (*Gloriosa superba*).

Richardson happily hunted and collected natural history for his other correspondents too. They were equally delighted to receive his specimens of northern plants and the presents of game that he sent to them. Robert Uvedale, for example, similarly received boxes containing Richardson's 'kind present of foul mosses [and] seeds all, in good condicon'. Uvedale's herbarium also contains the first specimen of Killarney fern (*Trichomanes speciosum*) that was collected by Richardson 'in the moist and shady rocks nigh Bingley'. The plant can still be found in the same locality today.

Other people in London with whom Richardson corresponded include James Petiver, Adam Buddle and the naturalist brothers William and James Sherard. Richardson also received letters from Philip Miller at the Chelsea Physic Garden and the botanist William Vernon in Essex (from Peterhouse College, University of Cambridge). Outside London Richardson corresponded with the antiquary Ralph Thoresby, who lived in Leeds, as well as the physician William Chambers in Hull. Similarly, the botanist in charge of the physic garden in Edinburgh, James Sutherland, as well as the curator there, Robert Wood, wrote to Richardson from Scotland. These individuals were not only corresponding with Richardson but with each other as well and they were exchanging all sorts of natural knowledge in the form of botanical news, publications and physical plant samples including seeds and dried specimens.

There was also much discussion about gardening practices and the difficulties of cultivating plants in Britain in the early eighteenth century. While enquiring after Richardson's garden, they also updated each other on how their own gardens were faring. In 1722 James Sherard thanked Richardson for his 'obliging Letter' but was 'very sorry' to hear that Richardson had 'had so ill success with [his] plants'.

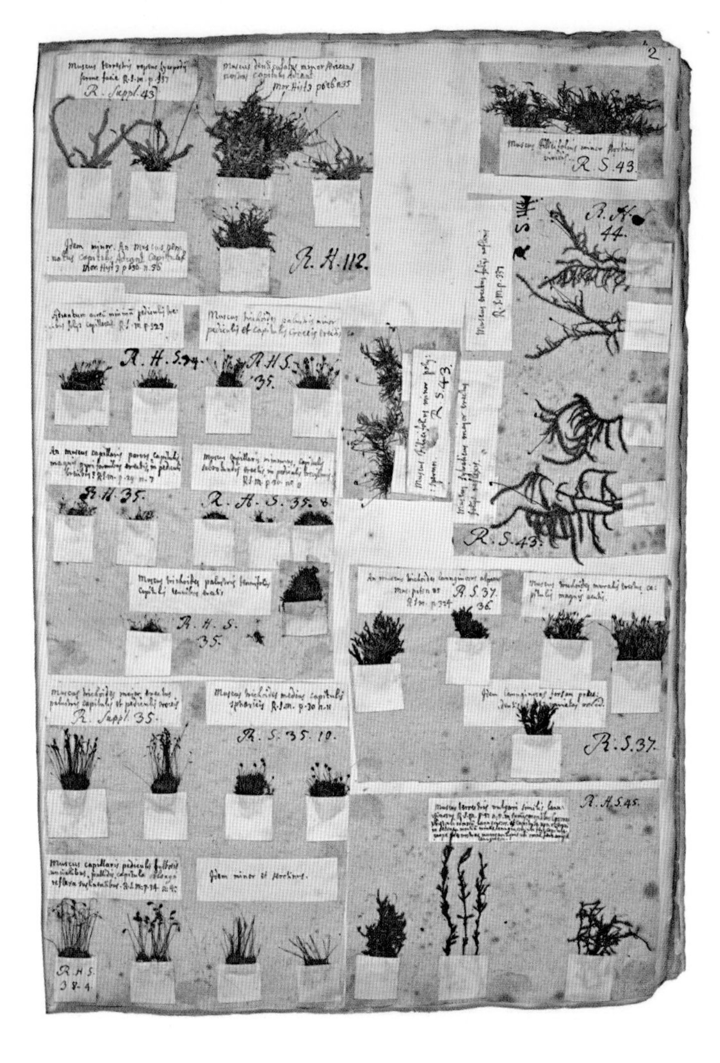

Left Richardson's love for natural history, and botany more broadly, is exemplified by the variety of plants found in his herbarium. The page shown here for example, shows a collection of bryophytes.

These men appear to have met frequently in person and often took 'pleasure' with 'friends' in 'looking over ye mosses' or whatever other natural history specimen had most recently been received. When they did they took the time to inform other friends, in writing, of their civilities to each other. For example, as Sherard wrote to Richardson, they drank to his 'good health in an evening after feasting on yr kind present'. In turn, this shaped collecting practices and some of the British and European samples found in Sloane's botanical collection mention or connect more than one contributor. For example, the sample of *Lloydia* was collected by Richardson while botanizing in North Wales with Edward Lhwyd.

It is evident that these individuals spoke to each other often, in writing and in person. These exchanges frequently feel like informal, but detailed, conversations about a mix of topics, from personal updates on health to the latest botanical news, gardening practices, and the cultivation of plants. They are an important reminder that many different people were involved in the movement of botanical material across Britain, and that it was not simply one connection between Sloane and a collector, in this instance Richardson.

Richardson's contemporaries considered him an influential figure in expanding the list of British species and identifying the distribution and habitats of British flora not least because of his sustained botanical investigations across the country. His research on mosses, in particular, came at a time when there was less understanding of mosses and lichens and his contribution to this area of botany should not be underestimated. He was perhaps the first Yorkshire naturalist to have collected lichens, and the plant genera *Richardia* L. and *Richardsonia* Kunth. were named in his honour.

From surviving correspondence as well as specimens it is also clear that Richardson played a central role in collecting and exchanging plant material across Britain. What makes the specimens from Richardson found across the Sloane Herbarium so interesting for historians as well as botanists, is their context. This is the context of the relationship that Richardson not only had with Sloane, but the connections that Richardson developed with so many other individuals as well. These connections give a sense that British and European botanical exchanges were more informal than those that brought material from further afield. They could, more routinely, be the product of face-to-face encounters as well as written correspondence, taking place in a variety of spaces and between multiple people.

BRUNO TOZZI

ADRIANO SOLDANO

Bruno Tozzi was an Italian botanist and mycologist, born in Florence in 1656. He became a Benedictine monk in 1676, joining the Order of Vallombrosa, an order that managed the vast Vallombrosa Forest in Tuscany from the eleventh until the nineteenth century. Tozzi's interests in natural history were wide ranging. He studied birds, insects and minerals, underpinned by precise observation and drawing, but botany – not only vascular plants, but also fungi, lichens, algae and bryophytes – was his main focus. Tozzi's motivation to study and collect plants might have come from the traditional commitment of the monks of his order to use natural plant-based remedies. In his manuscripts he certainly demonstrates an interest in this topic, for example concerning the use of aqueous extract of *Lythrum salicaria* for improved eyesight. Moreover, also resident at Vallombrosa at the same time as Tozzi was Virgilio Falugi whose botanical interests encompassed both the taxonomy and geographical distribution of plants. An interest in plants was not out of place for someone in monastic orders. For example, Paolo Boccone, a further eminent Italian naturalist and monk of the Cistercian order had recently completed a series of botanical explorations that had resulted in the identification of several new species.

Tozzi undertook botanical excursions during the period from 1699 to 1734. His various explorations were recorded in letters that he wrote to his brethren and fellow botanists Falugi and particularly Biagio Biagi, a fellow monk who was an expert in botany and who accompanied Tozzi in some excursions from 1699 to 1705. The great majority of the places he botanized were in Tuscany and he travelled widely

Below Tozzi's portrait was published in a commemoration of his achievements in the journal *Nuovo Giornale Botanico Italiano* in 1938.

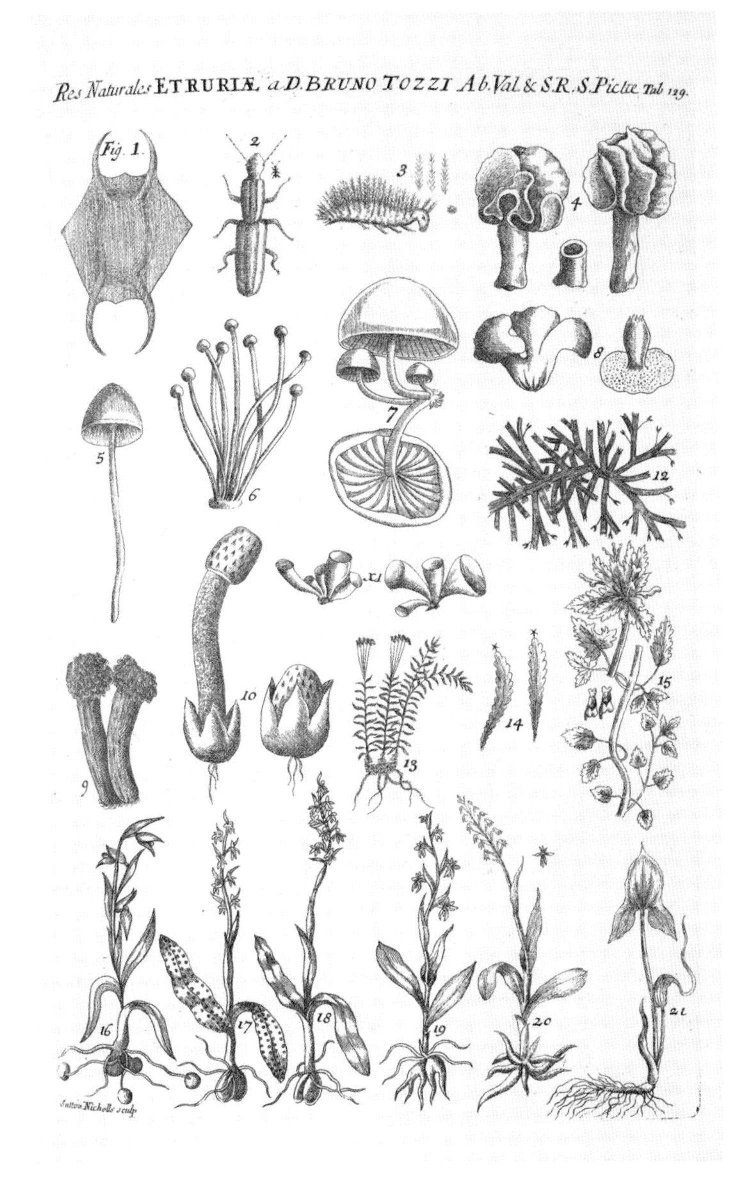

Left Petiver's table 129, published in 1767, contained animals, fungi and plants depicted by Tozzi.

across the region and its islands. He also collected plants more widely in Italy; northwards into Emilia Romagna and Venetum and south to Latium. His travels were made on foot or by river boats even though he was a skilled horseman. In the various places mentioned in the letters, Tozzi recorded not only the rarer plants but also more common species and his observations often constitute

Left *Silene lanuginosa* on the Pania Secca, Apuan Alps, one of the two possible localities from where Tozzi would have collected the specimen of this species in the Sloane Herbarium.

a baseline for the flora of the provinces that he traversed. Tozzi made a great collection of plants but few survive. One small volume, containing 27 sheets and just over a hundred samples, is preserved in the Botanical Museum of Florence. Further specimens are at Oxford in the herbarium of William Sherard who Tozzi had met in 1699 in Florence. There are also 36 specimens in the Sloane Herbarium.

The Tozzi specimens in the Sloane Herbarium today were originally in the herbarium of James Petiver and many have the collector ('Tozzi' or 'Br.Tozzi') indicated in Petiver's hand. After Sherard had met Tozzi in Florence, Tozzi subsequently sent him illustrations of fungi, orchids and rare plants. On learning this, Petiver also made contact with the Italian monk and from 1713 Tozzi and Petiver entered into correspondence (writing in Latin). Tozzi sent Petiver illustrations of Tuscan orchids and insects together with specimens of plants. Most of the specimens are flowering plants but they also include ferns and a bryophyte. The Sloane Herbarium also includes a list of 25 plants compiled by Tozzi and sent to Petiver, 22 of which correspond to specimens found in the herbarium today.

All of the Tozzi specimens in the Sloane Herbarium have labels with no indication of a more precise locality, except four that are from '*Etruria*' (i.e. Tuscany). However most are from Tuscany and Emilia Romagna and they can often be exactly located with reference to the accounts of Tozzi's collecting trips recorded in his letters that are preserved at the National Library of Florence and at the Royal Society Library of London. The collection dates range from 1701 to 1712 and the specimens come from a wide variety of habitats – from the Tuscan islands and from coastal areas, wetlands, hilly and rocky areas but predominantly mountainous meadows in the mainland. Some of the specimens in Sloane's herbarium are unique and are not duplicated in either Oxford or Florence.

Tozzi's collections in the Sloane Herbarium include species that are very restricted in their distribution such as *Silene lanuginosa* (fluffy campion), which occurs only in the Apuan Alps of northern Tuscany and specifically in the provinces of Lucca and Massa Carrara. We know from his letters that the sample in the Sloane Herbarium would have been collected either near the Passo delle Porchette in August 1703 or during the ascent of the 1,711 m high mountain Pania Secca in July 1704. In either case, Tozzi's specimen is the oldest surviving example of this species because there is no

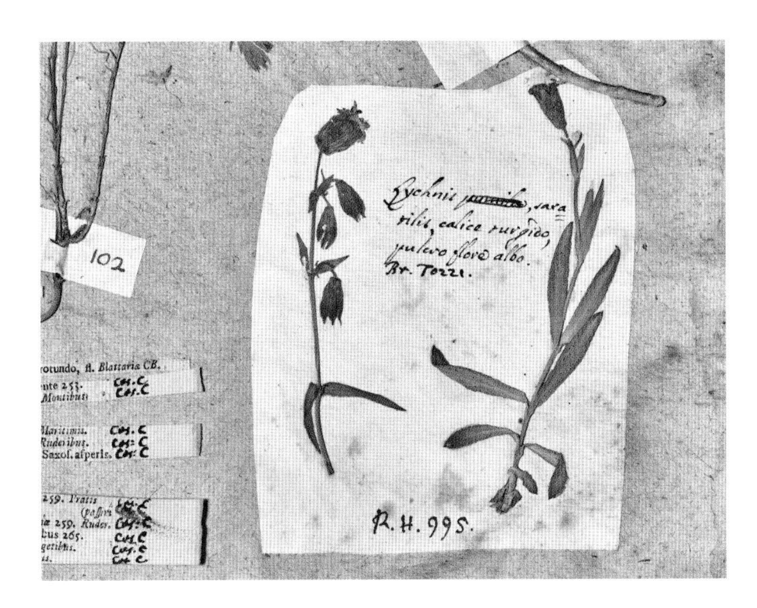

Left The *Silene lanuginosa* collection in the Sloane Herbarium made by Tozzi in the Apuan Alps.

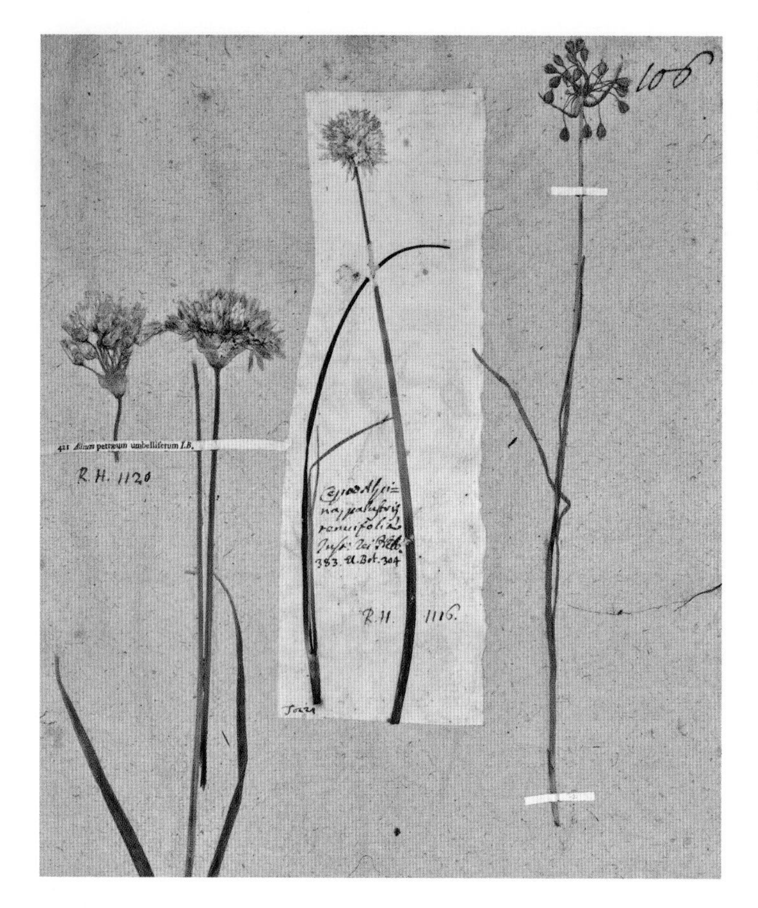

Left A specimen of chives (*Allium schoenoprasum* subsp. *schoenoprasum*; centre) collected by Tozzi near to Lago Santo in the Apennines. It is the earliest collection of chives from Italy.

material in Micheli's herbarium. Tozzi was accompanied on his 1704 expedition by the young botanist Pier Antonio Micheli (see p.32). Micheli accompanied Tozzi on a number of his early excursions in Tuscany and also on a much later excursion in 1730. Tozzi is credited with providing great support and encouragement to the young Micheli in the development of his botanical and mycological skills.

The polynomial name that Tozzi used for the aforementioned *Silene lanuginosa* on his label in the Sloane Herbarium was *Lychnis saxatilis calice turgido pulcro flore albo*. It was a name that remained unpublished. Indeed, Tozzi did not publish any of his findings. Micheli, who also collected a specimen of this species during their 1704 excursion published a different polynomial name (*Lychnis alpina saxatilis angustifolia alba caule & foliorum marginibus lanuginosis calyce amplo patulo*) in 1723.

In Emilia Romagna in 1704 Tozzi collected a specimen of chives (*Allium schoenoprasum* subsp. *schoenoprasum*) near Lago Santo, in the Apennines. It was the first collection for Italy as the first published record of it was not made until 1785 by Carlo Allioni. Indeed, many of the specimens collected by Tozzi provide the earliest records of species for at least the province in which they were collected. Tozzi's material was highly valued by Petiver. At a meeting of the Royal Society held in May 1714, Petiver proposed Tozzi as a Fellow and he was formally elected in November 1715.

In 1716 Petiver published a list of 10 plants 'sent me from the Reverend Bruno Tozzi which gathered in the Island Elba or Ilva' in his *Petiveriana seu Naturae Collectanea II.* Others were published by the Petiver in 1767, posthumously for both Tozzi (who died in 1743) and Petiver (who died in 1718), in his *Jacobi Petiveri Opera Historiam Naturalem Spectantia or Gazophylacium* (see image on p.107). This work included six plates of Tozzi's illustrations, one with orchids, the second with plants, fungi and animals and the other four with animals (and especially insects).

In 1719 Tozzi joined the Società Botanica Fiorentina that his friend and student Micheli and other botanical enthusiasts had founded in 1716. In 1729 Micheli dedicated to Tozzi the genus *Tozzia* that Carl Linnaeus mentioned in his *Species Plantarum.* Today the name is still in use for a group of hemiparasitic plants in the Orobanchaceae family.

Tozzi died in 1743, having risen to hold a prominent role in his monastic order, serving as its secretary general. As a botanist he did not directly publish anything but he was an important mentor and teacher. He is remembered today because his observations and specimens provide many first records for the Italian regions of Tuscany, Emilia Romagna and Latium; dozens of first records for all of Italy and several first records of species anywhere in existence.

JOAN SALVADOR I RIERA

NEUS IBÁÑEZ AND LAIA PORTET I CODINA

Joan Salvador i Riera was 20 years old when he became the correspondent of an English apothecary who was not only 20 years older than him but also a renowned promoter of the natural sciences. Demonstrator at the Chelsea Physic Garden and a member of the Royal Society, James Petiver was one of the most avid collectors of his time and a close collaborator of Sir Hans Sloane. Conversely, Salvador had just returned from studying chemistry and botany in France and his family's collection was at an embryonic stage. Yet, the letters and plant specimens that they exchanged from 1706 to Petiver's death in 1718 prove that the difference in age and professional status never became an impediment to their intellectual and material transactions. On the contrary, their correspondence was mutually beneficial and, as years passed by, Petiver grew to treat the younger and less experienced Salvador as an equal.

Below Portrait of Joan Salvador.

Opposite A plate in *Plants already engraved in Mr. Petiver's english herbal* dedicated to Salvador by Petiver.

Joan Salvador i Riera followed his father, Jaume Salvador i Pedrol, in developing interests in botany and natural history. He would become one of the greatest Spanish botanists of his century. Using his father's contacts, he became a disciple of Pierre Magnol at the Jardin des Plantes in Montpellier and a pupil of Joseph Pitton de Tournefort at the Jardin du Roi in Paris. He also became acquainted with the eminent Italian naturalists Michelangelo Tilli and Giovanni Battista Triumfetti when the outbreak of the Succession War in Spain forced him to leave France – an enemy country – and spend a few months in Italy before being able to return home in

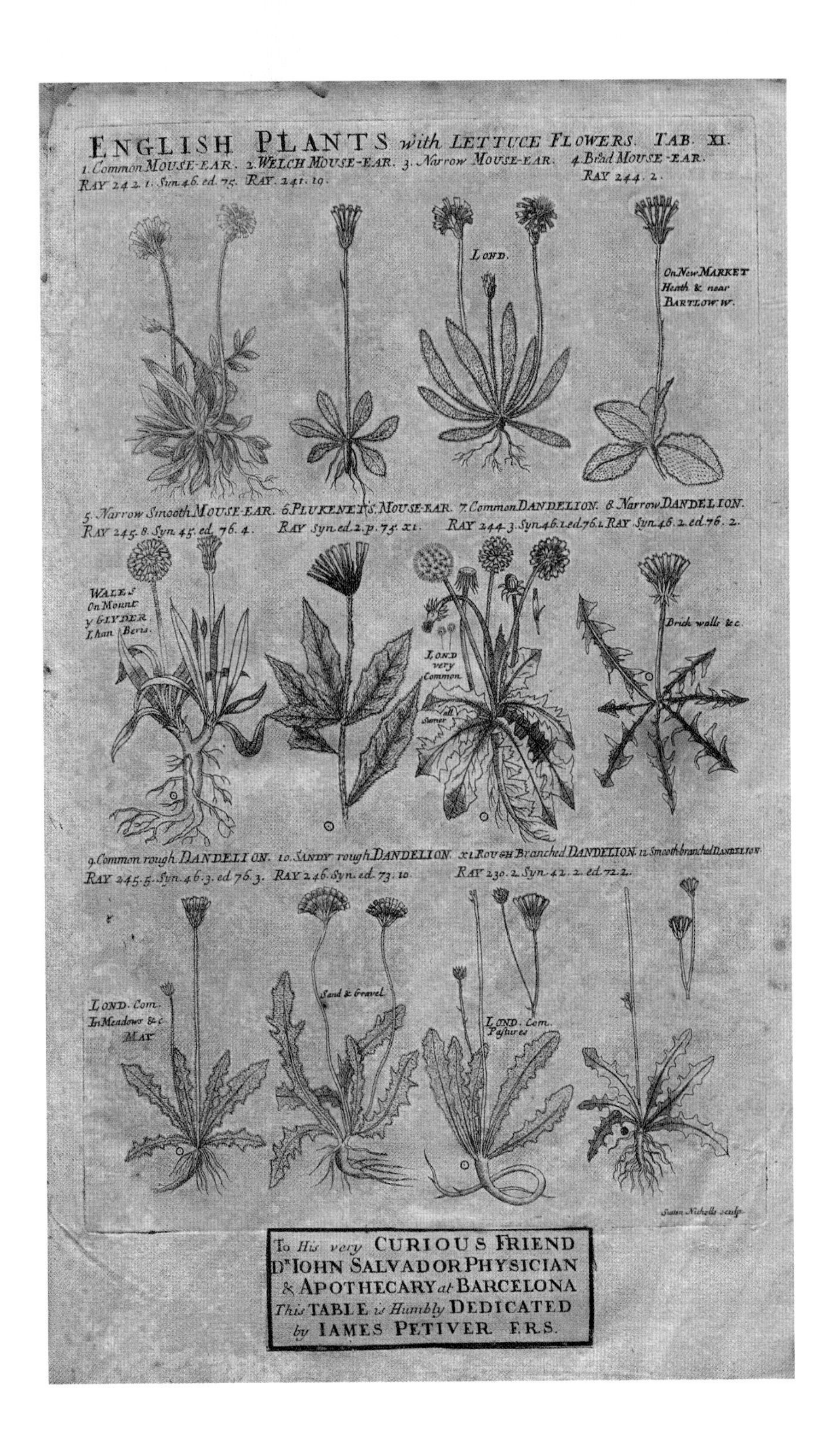
ENGLISH PLANTS with LETTUCE FLOWERS. TAB. XI.
1. Common MOUSE-EAR. 2. WELCH MOUSE-EAR. 3. Narrow MOUSE-EAR. 4. Broad MOUSE-EAR.
RAY 242. 1. Syn. 46. ed. 75. RAY. 241. 19.
RAY 244. 2.
LOND.
On New MARKET Heath & near BARTLOW. W.
5. Narrow Smooth MOUSE-EAR. 6. PLUKENET'S MOUSE-EAR. 7. Common DANDELION. 8. Narrow DANDELION.
RAY 245. 8. Syn. 45. ed. 76. 4. RAY Syn. ed. 2. p. 75. xi. RAY 244. 3. Syn. 46. 1. ed. 76. 1. RAY Syn. 46. 2. ed. 76. 2.
WALES On Mount y GLYDER Lhan Beris.
LOND very Common.
Brick walls &c.
9. Common rough DANDELION. 10. SANDY rough DANDELION. 11. ROUGH Branched DANDELION. 12. Smooth branched DANDELION.
RAY 245. 5. Syn. 46. 3. ed. 76. 3. RAY 246. Syn. ed. 73. 10. RAY 230. 2. Syn. 42. 2. ed. 72. 2.
LOND. Com. In Meadows &c. MAY
Sand & Gravel.
LOND. Com. Pastures
To His very CURIOUS FRIEND
Dr IOHN SALVADOR PHYSICIAN
& APOTHECARY at BARCELONA
This TABLE is Humbly DEDICATED
by IAMES PETIVER F.R.S.

the autumn of 1706. The Spanish succession conflict, which pitched Habsburg Spain and its allies, the English, Dutch and Holy Roman Empire against Bourbon Spain and France in a war over who would succeed the childless Charles II of Spain, froze Salvador's relationship with French scholars. He resumed them as soon as the war was over and in 1715 he became a correspondent of the Parisian Academie Royale des Sciences and was later invited by Antoine de Jussieu, Director of the Jardin des Plantes in Paris (and who had been a fellow student with Salvador), to botanize around Spain and Portugal in the same fashion his father had done with Tournefort in the early 1680s.

Confined to Barcelona and its surroundings for the duration of the war and deprived of news from France, Salvador invested the years between 1706 and 1714 in the collection of curiosities and the herbarium that his grandfather, Joan Salvador i Boscà, and his father, Jaume Salvador i Pedrol, had started. The Spanish War of Succession, despite all its downsides, encouraged the expansion of Salvador's scientific network with new contacts from England, the Low Countries and the Holy Roman Empire. In 1711 these relationships led him to conduct the first comprehensive botanical exploration of Mallorca and Menorca with the English surgeons George Boucher and James Campbell who were both resident in Mahon, the main English garrison in Menorca. It was also during wartime when Salvador sent his first letter to his soon-to-be most fruitful correspondent, the respected London apothecary, James Petiver. Their professional relationship enriched both of their herbaria and resulted in several publications in *Petiveriana Naturae Collectanea*.

As it stands nowadays in the Botanic Institute of Barcelona,

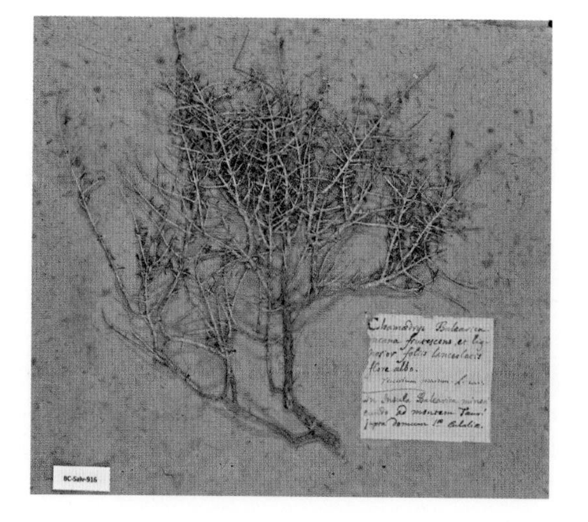

Below The type specimen of *Teucrium subspinosum* in the Salvador herbarium. The species was described by German botanist Willdenow in 1809, nearly 100 years after it was collected by Salvador.

Above The Salvador Collection in the Botanic Institute of Barcelona, Spain.

the Salvador Collection is the most complete cabinet of curiosities preserved in Europe and the most authentic reproduction of how it looked in the early modern period. It is composed of a library of 1,548 volumes and a collection of 9,237 specimens, housed in the tasteful eighteenth-century set of furniture that was built under the orders of Josep Salvador i Riera, Joan's younger brother. This cornucopia of materials is complemented by three centuries of private correspondence and family papers, and the outstanding herbarium.

Salvador's herbarium consists of 4,931 specimens and is the oldest in Spain. To document the species in the herbarium, Salvador used the family library alongside books and information sent to him by friends and correspondents and it is classified using Tournefort's taxonomic system. For the collection of specimens, Salvador relied on his own travels and his family's network of correspondents.

Salvador's herbarium was enriched with an annex that included American plants sent by James Petiver. Petiver dedicated an engraving to Salvador and in return for the specimens he sent to Salvador, Petiver received more than a thousand specimens from Salvador that are today in the Sloane Herbarium. They are identifiable by the signature *Salv* and his handwriting. Salvador had collected the bulk of them during his eight-month-long journey around Spain and Portugal between 1716 and 1717 accompanied by Antoine et Bernard de Jussieu. Around 50% are not localized but the labels indicate that 372 sheets were collected in Spain, in the region of Pyrenees, in Valencia and Catalonia, with many plants collected around the localities of Barcelona and Montserrat, a mountain range near to the city. Eight plants were collected in Portugal. The other 143 sheets correspond to the Balearic plants that Salvador obtained from his trip to Mallorca and Menorca in 1711.

Opposite A specimen of *Atropa belladona* (top) collected by Salvador in Petiver's collection. The 'Salv' written in the bottom left corner of the label indicates its collector is Salvador.

Salvador's material contributed species to Petiver's and subsequently Sloane's herbarium that were not otherwise represented in their collections. Salvador described three new taxa from Montserrat in *Petiveriana Naturae Collectanea* II (Petiver, 1716), specimens of which are in the Sloane Herbarium. His botanical expedition to the Balearic Islands enabled him to name 26 new taxa, 16 of which were later published in *Petiveriana Naturae Collectanea* III. Salvador wrote a manuscript on Catalan botany called *Botanomasticon catalonicum* that was never published and has since been lost. The publications in *Petiveriana Naturae Collectanea* and his herbarium specimens are therefore the best proof today of his calibre as a botanist.

None of the descriptions proposed in *Petiveriana Naturae Collectanea* were used by Carl Linnaeus when he was preparing *Species Plantarum*, the basis for modern binomial system for naming of plants. This is perhaps due to the limited distribution of this work following the death of Petiver one year after its publication in 1717. Consequently, and despite their importance, Linnaeus used a list of Balearic plants sent to him by Antoine Richard, currently lost, rather than the Salvador list, in the preparation of *Species Plantarum*.

Nevertheless, Salvador's collections from regions in Iberia and the Balearic Islands that were not otherwise well explored at that time have remained not only historically but also scientifically important. One species based on Salvador's material was described by the German botanist Carl Ludwig von Willdenow in 1809. Another species collected by Salvador in the Balearic Islands was recognized as a new species, *Teucrium cossonii*, in 1972, nearly 200 years after his death.

152
RH. S.
360.
In regno Valentino, et murciano.

COLLECTORS IN THE FIELD: THE ATLANTIC

MARK CARINE AND ROBERT HUXLEY

The 'Atlantic World' was where Sloane made his most extensive and most significant field collections. Written up in his monumental and richly illustrated *A voyage to the islands Madera, Barbados, Nieves, S. Christophers and Jamaica*, the seven volumes of plants that Sloane collected as a young man on that Atlantic voyage constitute the first seven volumes of the Sloane Herbarium today. In many respects they can be seen as its foundation.

For many historians it is the relationships and interactions between the peoples of the Americas, the Caribbean, Africa and Europe that are of particular interest in the Atlantic World. Key themes include colonialism, trade and slavery and those themes are certainly pertinent to understanding this region's collections in the Sloane Herbarium. Edward Bartar, the subject of our first essay, is a case in point. Bartar was a West African slave agent based at Cape Coast Castle in modern-day Ghana, then the African headquarters of the Royal African Company. Founded in 1660, initially to trade in gold, the Royal African Company was soon engaged in the transatlantic slave trade. Bartar was certainly not alone in his links to slavery among West African collectors. Where precise collecting locations are given for West African specimens they are invariably slave ports, and it is this transatlantic slave trade that is the reason why plants from the region are in the Sloane Herbarium today.

Links to slavery are also evident in collections from the Americas and the Caribbean. There are specimens from Cartagena in Colombia and from Portobelo in Panama. Both were ports where British ships traded in enslaved people with the Spanish who did

Opposite An airplant, *Tillandsia polystachia*, collected by William Houstoun.

Left The map of the Atlantic and of Jamaica from Sloane's *Natural History of Jamaica*.

not have colonies in Africa. Contributors from English colonies with plantation economies are also prominent: colonies such as South Carolina, Maryland and Virginia in North America; and Barbados, Antigua and Jamaica in the Caribbean. Not all collectors would have been directly involved in slavery, but they were all living – and collecting – in slave societies. Slavery would have been part of their normal lives. In his account of Jamaica, Sloane himself recounts, 'I saw in this harbour and Bay [Port Royal] a Ship come from Guinea, loaded with Blacks to sell. The Ship was very nasty with so many People on Board.' In this chapter we include our essay on Sloane, its main focus being on his collecting in Jamaica.

Colonial expansion – and conflict – in the Atlantic World is also reflected in the Sloane Herbarium. Unsurprisingly, it is the English colonies that are best represented but plants from regions colonized by other European powers are also to be found. There are, for example, collections made in Darién, Panama during the unsuccessful attempt by Scotland to establish a colony providing an overland trade route connecting the Pacific and Atlantic. There is also a fern specimen from the French island of Martinique, contributed by the English ship's surgeon James Stewart. The label records that it was 'Taken in a French prize'. This would have been a ship seized during the War of the Spanish Succession, that pitched England and France on opposing sides; French ships – and the botanical specimens they contained – would have been fair game.

English colonial officials were among those contributing specimens. John Lawson, who rose to become Surveyor General of the Colony of Carolina, was a colonial official who made a significant contribution to the Sloane Herbarium. Lawson was the author of a celebrated account of his travels through Carolina and contributed nearly 300 specimens before he succumbed to an untimely and painful death.

If trade, colonialism and slavery provided the infrastructure for collecting in the Atlantic World, then demand for exotic new plants was an important stimulus for that collecting. The demand led to dedicated expeditions and created opportunities for trade. Encouraged by James Petiver, David Krieg visited Maryland in 1698 and contributed some of the first specimens to be collected in the colony. Subsequently, William Houstoun and Mark Catesby were sponsored by Sloane and his contemporaries to collect in what is now the southeastern United States and the Caribbean. Many of these collectors were employed to gather information on the 'natural wealth' of the new colonies, recording information on actual and potential uses of the fauna and flora that they encountered. Our final contributor is John Bartram. Unlike the other plant hunters, he was American born rather than a traveller from Europe. He was also able to turn his botanical explorations into a successful business, helping to satisfy the demand from naturalists and gardeners in Europe for new and interesting plants from the Atlantic World.

Below A folio of specimens collected in 1733 by James Oglethorpe, the social reformer and founder of the colony (now state) of Georgia. The specimens were collected in the same year that he established the colony.

EDWARD BARTAR

KATHLEEN S. MURPHY

In the late seventeenth century, European presence along the West African coast was confined to a series of trading forts, known as factories. The largest, best fortified, and most important of the English factories was Cape Coast Castle, in what today is southern Ghana. There the English Royal African Company made its African headquarters. Officially at least, all English trade with West Africa in this period was monopolized by the Royal African Company. The castle's commander oversaw the Company's trade in enslaved Africans and West African natural commodities. Cape Coast Castle's halls served as home for the dozens of white and black traders, artisans, translators, clerks, soldiers, domestic servants and enslaved Africans employed by the Company, while its dungeons confined hundreds of captive Africans bound for the New World. In a contemporary view of the castle, on the adjoining page, the English trading fort is depicted as an imposing presence over the surrounding countryside, and featuring bastions, canons and a remarkably large English flag prominently flying. Just to the right of the castle lies a second fortified structure. In many ways it was a miniature version of the first, complete with canons, fortified walls and a large English flag. This, however, was not one of the Royal African Company's factories. Instead, it was the home of Edward Bartar. Bartar was a slaving agent, employee of the Royal African Company, and collector of natural history specimens that found their final home in the Sloane Herbarium.

Bartar was of Anglo-African descent; most likely his mother was a local African woman and his father was an employee of the Royal African Company. This connection to the Royal African Company helps to explain why the Company paid to have Bartar educated in England in the early 1690s. During his time in England, the Anglo-African became acquainted with English natural history

Above The fortified home of slave trader and collector Edward Bartar (centre) appears like a miniature version of Cape Coast Castle (left) in this seventeenth-century engraving.

collectors and botanists including Leonard Plukenet, Samuel Doody and James Petiver. Bartar returned to the Gold Coast in 1693 as an employee of the Royal African Company, drawing an annual salary of £30. His responsibilities with the Company were many and took him throughout the Gold Coast. Bartar collected the Company's debts, purchased corn and cattle, delivered messages and goods between Cape Coast Castle and the Company's factories, attempted to negotiate peace with local African rulers, assisted in the defence of factories when such attempts failed, and transported enslaved Africans to the castle.

In time Bartar came to be one of the most powerful men along the Gold Coast. His contemporary Willem Bosman, who was stationed at one of the neighbouring Dutch factories, declared that Bartar 'hath a greater Power on the Coast, than all the three English Agents together (in whom the chief Command of the Coast is vested jointly)'. Bosman explained that the agents' stay on the coast was relatively brief, whereas Bartar brought many years of experience, extensive connections among local African rulers, and command over a significant personal army. The Royal African

Company's agents relied on Bartar's advice and assistance even though he was no longer employed by the company by 1700. According to Bosman, 'whoever designs to Trade with the English, must stand well with him before he can succeed'. The arrival of a commander at Cape Coast Castle who was determined to assert his predominance over English trading led Bartar into conflict with the Royal African Company. He spent the last two years of his life as a fugitive living first at the Dutch factory El Mina and later at the court of a local African ruler.

Bartar's friendship with Petiver, Plukenet, Doody and other English naturalists seems to have been the primary motivation for his collection of natural history specimens in the years immediately following his return to West Africa in 1693. Shortly after Bartar's departure from England, Petiver wrote to urge him to collect. The English naturalist reminded Bartar that his friends eagerly awaited the many specimens he was sure to send and that once a week they 'remember you in a Glass of Nottingham Ale'. The slaving agent sent collections of plants, butterflies, shells, 'some Medicinal Roots' and 'an elegant hairy Catterpillar'. Petiver thanked Bartar in print for sending '3 or 4 Quires of Plants' including what the naturalist described as the matice-weed, malabar bindweed, scorpion sena and trefoil ground-bean. Among plants Bartar collected were many that Petiver declared 'strangers and curious', including an African willow in flower that the naturalist described as 'one of the strangest plants I have seen'. In return, Petiver sent Bartar gifts such as newspapers and English beer, delivered

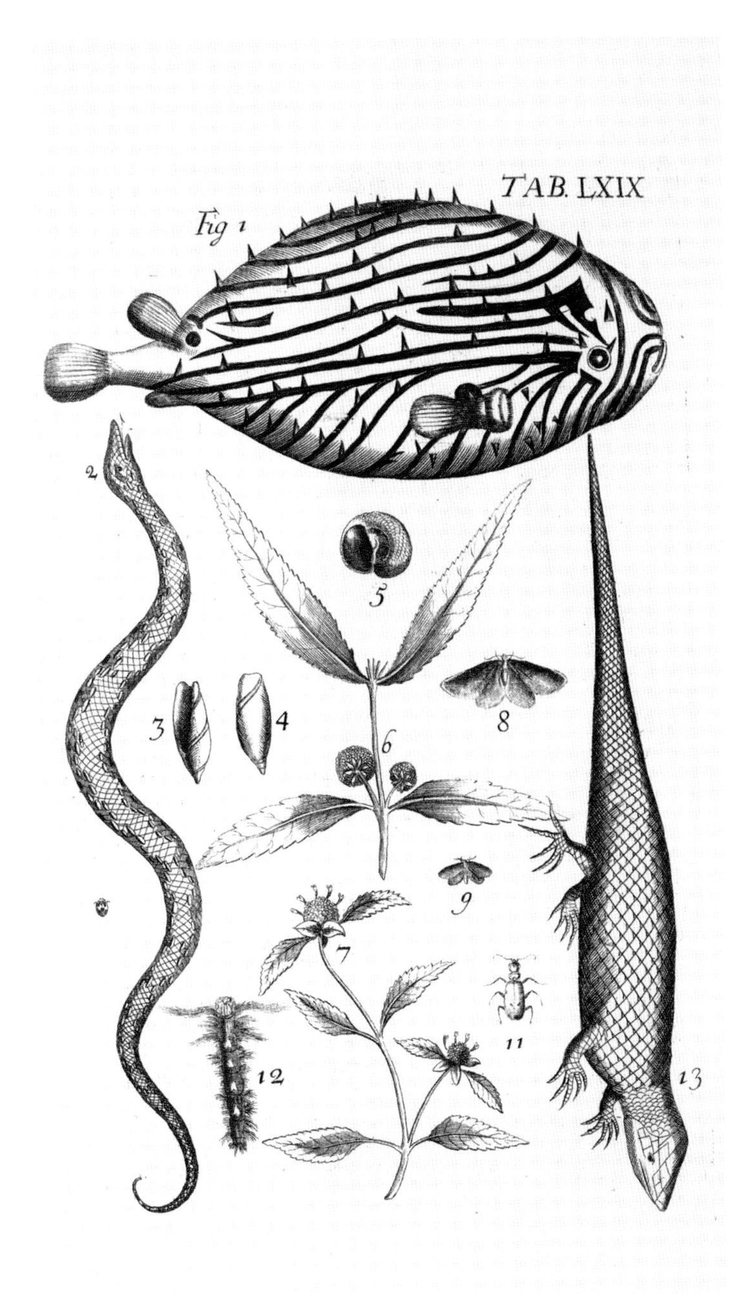

Opposite This specimen bears James Petiver's printed description of the plant as the 'Amourettes *Guineensis* major, paniculâ speciosiore' gathered '*about* Cape-coast in Guinea' by the slave trader Edward Bartar.

Left James Petiver featured the West African 'elegant hairy Catterpillar' (fig. 12) he received from Bartar in his *Gazophylacii Naturae & Artis* (1711).

letters to Bartar's wife in England and undertook other favours in the metropolis on the slaving agent's behalf.

Despite a promising start, Bartar's collecting efforts ultimately did not meet his English friends' high expectations. His specimens were often incomplete, missing either the flower or seed. Further, the slaving agent collected little after his first few years back on the Gold Coast. Eager for additional specimens, Petiver urged Bartar to employ his enslaved Africans as collectors and reminded him of the frequent opportunities he had to send specimens by way of the transatlantic slave ships that called at Cape Coast Castle. 'I could never have thought,' Petiver declared in 1697, that Bartar would have allowed 'my very kind friends Mr. Wingfield and Mr. Ric[hard] Planer,' to leave Cape Coast without any specimens. 'What was worse,' Petiver declared, was that Bartar prevented the two slave ship surgeons from collecting themselves by promising that he would do so and denying them the assistance of his enslaved Africans.

Bartar, Wingfield and Planer were just three of the scores of individuals engaged in the transatlantic slave trade who also collected and transported natural historical specimens, including many now in the Sloane Herbarium. Nearly a quarter of Petiver's collectors within the lands touching the Atlantic Ocean, i.e. the Caribbean, South and North America, Europe, Africa and the Atlantic islands, were engaged in the slave trade. Some, like Bartar, worked as slaving agents at factories in West Africa or in the Americas. More frequently, slaving mariners collected specimens, transported the collections of others and recruited new collectors for metropolitan naturalists. Slave ship captains and surgeons like Wingfield and Planer gathered specimens along the West African coast, as well as in American ports where captive Africans were forced to disembark. Such collecting practices help to explain the presence of specimens from West African slaving ports frequented by British slave traders, such as Angola, Calabar, Cape Coast, Cape Three Points, Mount Serrado and Whydah, within the Sloane Herbarium.

The West African plants Bartar gathered in the 1690s initially formed part of Petiver's collection. After Petiver's death, Sir Hans Sloane purchased Petiver's museum, thus adding Bartar's plants to the Sloane Herbarium. Over a dozen specimens collected by Bartar along the Gold Coast can today be found in the Sloane Herbarium. Some are incomplete, like a specimen comprising two leaves labelled with a note indicating that their juice is good for 'sore eyes'. Such examples reflect Petiver's complaint that some of Bartar's specimens contained 'only leaves and [were] wanting either flower or fruit or both'. The significance of the specimens collected by Bartar thus lies more in the means by which they were collected than in the particular objects themselves. Along with specimens gathered by slaving mariners such as Wingfield and Planer, Bartar's collections testify to the ways in which British naturalists utilized the routes of commerce, including the inhuman commerce in enslaved Africans, in order to obtain the flora and fauna that formed the basis of their scientific study.

Above The label for this fragmentary West African specimen, likely written by Edward Bartar, reports that the 'The juse of this leaf is good for sore eyes the name unknown.'

SIR HANS SLOANE

JAMES DELBOURGO

Sir Hans Sloane is remembered as the posthumous founder of the British Museum (1759), which was created to house his vast collections of *naturalia* and *artificialia*. His herbarium was a focal part of these collections, numbering 334 volumes of plants from around the world, collected by many different travellers (see the other essays in this book). Sloane was both a collector of collections, amassing this herbarium through gifts, exchanges and acquisitions from his network of correspondents, and also a field botanist in his own right. At the core of the Natural History Museum's herbarium are the specimens collected by Sloane himself. These include specimens gathered in England and France, but his most significant field collecting was in Jamaica. The first seven volumes of his herbarium contain approximately 800 plants that Sloane collected from Jamaica. These were transferred from the British Museum to the Natural History Museum as the latter's *de facto* founding collection when it was established in 1881.

Sloane was born in Killyleagh, Ulster, as a member of the minority Protestant elite, which colonized the island and subjected its Catholic population to colonial rule. After a severe childhood illness, he sought his fortune as a physician in London, where he was assisted by Robert Boyle and other leading figures in science and medicine such as Thomas Sydenham, John Ray and John Locke. Sloane studied botany, horticulture, chemistry, surgery and medicine at venues such as Chelsea Physick Garden, as well as in France under Joseph Pitton de Tournefort at the Jardin du Roi in Paris and Pierre Magnol at the botanical garden in Montpellier, earning his MD from the University of Orange in 1683. After returning to London, he embarked on a more ambitious and hazardous journey: to Jamaica, where he served as physician to the governor during 1687–1689 and where he made extensive collections of plants, animals and curiosities.

Above Sir Hans Sloane portrayed by Stephen Slaughter in 1736 as President of the Royal Society.

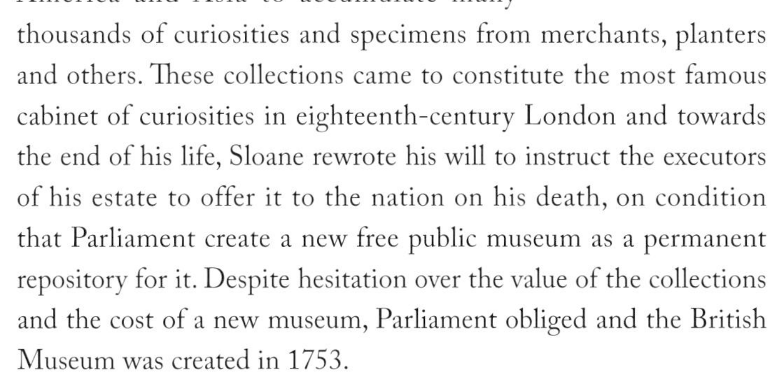

On returning to London in 1689, Sloane became one of the city's wealthiest physicians, attending the royal family and numerous aristocrats, becoming President of the Royal College of Physicians and the Royal Society concurrently. Thanks to the wealth of professional and intellectual connections he amassed, Sloane developed a worldwide correspondence, using the expansion of the British Empire in both America and Asia to accumulate many thousands of curiosities and specimens from merchants, planters and others. These collections came to constitute the most famous cabinet of curiosities in eighteenth-century London and towards the end of his life, Sloane rewrote his will to instruct the executors of his estate to offer it to the nation on his death, on condition that Parliament create a new free public museum as a permanent repository for it. Despite hesitation over the value of the collections and the cost of a new museum, Parliament obliged and the British Museum was created in 1753.

The essential context for Sloane's botanizing in Jamaica is Atlantic slavery. After the English seized Jamaica from the Spanish in 1655, they used it as a pirate base from which to plunder Spanish treasure fleets. Towards the end of the century, however, when Sloane worked in Jamaica, the planter interest came to dominate the buccaneering faction and, under the Royal African Company, the rate of slave importation from territories now encompassed by Ghana and Côte d'Ivoire began its steep climb, destined to continue throughout the eighteenth century. Between 1670 and Sloane's arrival in 1687, the number of Jamaican plantations jumped from 146 to 690. By 1700, British domestic consumption of sugar (including Barbadian sugar) had trebled, reaching 14,000 tons annually.

Sloane's motivations as a collector in Jamaica were scientific, medicinal, commercial and imperialist. He aimed to enrich himself as a plantation doctor while hoping to find new drugs unknown to the European market, like the anti-malarial Peruvian bark, then largely monopolized by the Spanish. As a learned naturalist and colleague of Ray and Tournefort, he aimed to catalogue plants and contribute to European taxonomy on a worldwide scale. His accounts of flora involved extensive assessments of their medicinal uses by locals and attention to their prices as commodities in the Americas and Europe. These assessments drew on his quizzing of the island's enslaved African population, whom he regarded as a living link to otherwise-lost knowledge of the Spanish and the indigenous Taíno people of the island. In Sloane's day there was, of course, no equivalent of the Nagoya Protocol of 2014, with its legal provisions that aim at equitable access to genetic resources and traditional knowledge, and the sharing of benefits arising from their use with local stakeholders. In this respect, like virtually all Europeans of his era, Sloane was a devout bio-prospector *avant la lettre*, if not a bio-pirate, searching out lucrative plant commodities for their commercial value. He hoped his new-found plants would benefit Protestant England in its struggle against rival Catholic powers Spain and France, even as he made his findings available to the 'Republic of Letters' (the international community of scholars across Europe and the Americas) by publishing a lavishly illustrated *Natural History of Jamaica* (1707–1725).

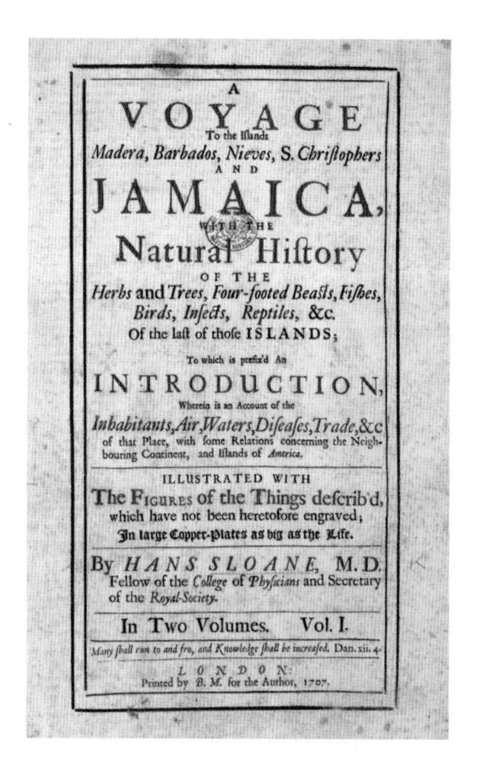
A
VOYAGE
To the Iſlands
Madera, Barbados, Nieves, S. Chriſtophers
AND
JAMAICA,
WITH THE
Natural Hiſtory
OF THE
Herbs and *Trees, Four-footed Beaſts, Fiſhes, Birds, Inſects, Reptiles,* &c.
Of the laſt of thoſe ISLANDS;
To which is prefix'd An
INTRODUCTION,
Wherein is an Account of the
Inhabitants, Air, Waters, Diſeaſes, Trade, &c
of that Place, with ſome Relations concerning the Neighbouring Continent, and Iſlands of *America*.

ILLUSTRATED WITH
The FIGURES of the Things deſcrib'd, which have not been heretofore engraved;
In large Copper-Plates as big as the Life.

By *HANS SLOANE*, M.D.
Fellow of the *College* of *Phyſicians* and Secretary of the *Royal-Society*.

In Two Volumes. Vol. I.

Many ſhall run to and fro, and Knowledge ſhall be increaſed. Dan. xii. 4.

LONDON:
Printed by *B. M.* for the Author, 1707.

Below Title-page of Sloane's *Natural History of Jamaica*, vol. 1, London (1707). Sloane's natural history was an encyclopaedic description of the plants, animals, peoples and commodities of Jamaica.

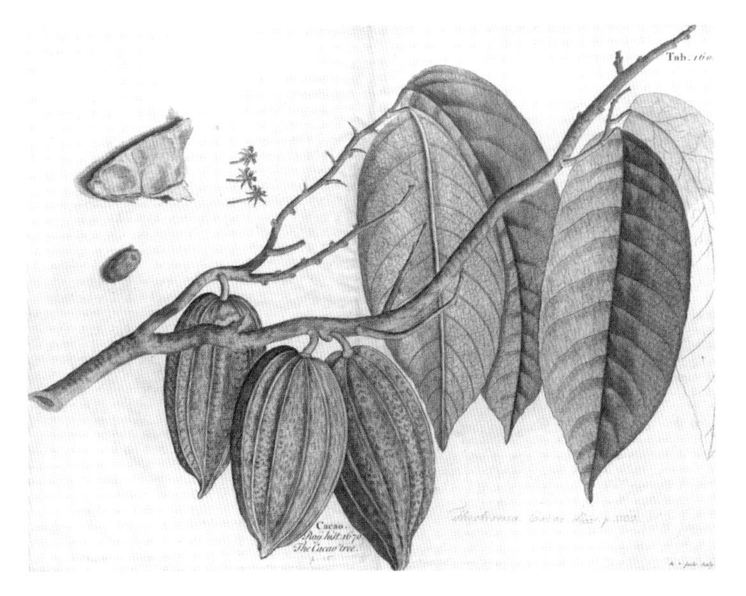

Left Cacao engraving from Sloane's *Natural History of Jamaica*, vol. 2, London (1725). Sloane's natural history included highly accurate life-size species drawings based on both living and dried specimens.

Sloane's treatment of his specimen of cacao (*Theobroma cacao*), the source of cocoa beans, exemplifies the pattern of multiple valuation that characterized his work in natural history. Cacao forms the basis of a detailed and highly accurate engraving in his *Natural History*, made by a combination of artists on both sides of the Atlantic. The Reverend Garrett Moore drew living plants for Sloane in Jamaica, and the Dutch draughtsman Everhardus Kickius inked over Moore's originals back in London over a decade later (1699–1701). Kickius combined these with sketches of the dried specimens Sloane had brought back, to produce pictures of plants that combined multiple diagnostic characters, as demanded by Ray's taxonomic system and to which Sloane adhered. Sloane compiled the textual *historia* of cacao during the years following his return from Jamaica, which he included with the finished engraving in Volume Two of his *Natural History*. This *historia* attends precisely to the anatomy of the species, but concentrates mainly on its use in recipes for making chocolate, the use of cacao nuts as money in indigenous American societies (which Sloane found highly curious) and the value of both nuts and chocolate in transatlantic markets.

Above Cacao specimen, likely harvested by enslaved Africans in Jamaica, with sketch of a living specimen done for Sloane by Everhardus Kickius.

Sloane not only talked at length with enslaved Africans about Jamaican flora but visited their provision grounds, where they grew their own food, often of African provenance. His specimen of *Sorghum*, the Guinea corn, still in his herbarium, was probably taken from one of these 'slave gardens'. How he acquired it, we cannot say with certainty. He may have stolen it; he may have exchanged for it; he may have paid for it; or he may have threatened violence (when he thought they were feigning illness, Sloane did sometimes threaten slaves with violence to force them back to work). On the other hand, we know that his associate, the apothecary James Petiver, whose herbarium Sloane acquired, arranged to have slaves paid a small amount of money to collect specimens for him. Many of the specimens Sloane took in Jamaica grew wild; some of the more common crops he sampled we know were harvested by slaves, including cacao. Ultimately, Sloane's Jamaica herbarium is an artefact of the institution of slavery that underpinned the island's colonial economy: whatever exchanges or payments may have procured plants must be understood in that context. Indeed, Sloane witnessed violence against enslaved men and women and condoned it, though this is unsurprising because of his marriage to Elizabeth Langley Rose, income from whose sugar estates he enjoyed for years and which supported his collecting. Friendships with planters afforded him shelter as he travelled in Jamaica and furnished critical advice on the island's natural history.

Sloane's Jamaica herbarium therefore allows us a valuable and rare glimpse into the multi-stage process by which pre-Linnaean scientific illustration worked at the turn of the eighteenth century, allowing us to analyze the compositional progression from specimen to sketch to published engraving. In his *Species Plantarum* (1753), Carl Linnaeus later advised readers that Sloane's pictures of species like *Theobroma cacao* were among the most accurate and reliable any naturalist could consult. Subsequently, under the modern system of plant typification established in the nineteenth century, a number of Sloane's specimens have come to act as taxonomic voucher material. In addition, much remains to be achieved through a systematic study of Sloane's herbarium as an index of historic species distribution. But the value of this herbarium is undeniably also as an artefact of interactions between enslaved African men and women and the English colonizers of the Caribbean during a decisive period of world history. It demonstrates the scientific productivity of slavery and imperialism during the long history of the Columbian Exchange that remade Europe, the Americas and Africa ecologically and economically in the early modern era. As such, it raises fundamental and challenging questions about whose labour and knowledge it contains, how it was produced, and how to confront its legacy today.

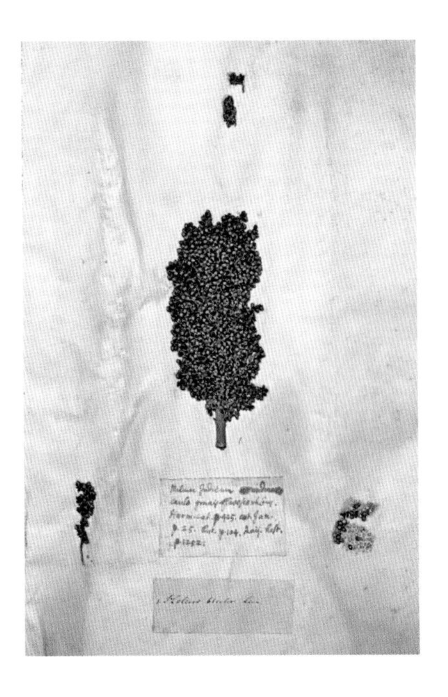

Below Guinea corn (*Sorghum*) specimen, used to feed slaves on the Middle Passage (the crossing from Africa to the Americas) collected by Sloane, probably from enslaved provision grounds in Jamaica.

JOHN LAWSON

ROBERT HUXLEY

Opposite Specimen of the sourwood or sorrel-tree (*Oxydendrum arboretum*) collected by John Lawson in North Carolina (centre top). A note in Lawson's writing has been added with name, a reference to '*a New Voyage*', location and date of collection and a collector's number.

Collecting natural history specimens in the seventeenth century could (and still can) be a perilous pursuit. Collectors risked shipwreck and disease, and encounters with local people could be violent. Perhaps the most tragic of these encounters led to the untimely death of one of the most talented North American collectors of the early eighteenth century. While exploring in North Carolina, botanist and surveyor John Lawson, along with Swiss nobleman Baron von Graffenried and their enslaved African bearer, whose name we do not know, were captured by the Tuscarora tribe and tried for crimes against their people. Von Graffenried was released to tell the tale but Lawson was tortured and killed. Their bearer's fate is unknown. During his time in Carolina, Lawson had observed, described and recorded the native fauna and flora and the customs of the people he encountered. Accounts of his expeditions were eloquently recorded and published as *A New Voyage to Carolina*, now regarded as a classic of early American literature. Some 300 of the plants that he collected in North America can be found distributed through several volumes of Sir Hans Sloane's herbarium.

Lawson was probably born in London around 1674, the son of Dr John Lawson and his wife Isabella Love, a wealthy family with lands in Yorkshire. John may have spent time and received his early education there. On his return to London he attended lectures at Gresham College near his family home. Founded in 1597, the college is regarded as the first English university outside Oxford and Cambridge and in the mid-seventeenth century it had become the meeting place for the fledgling and as yet unnamed Royal Society. Lawson was always keen to make an impression on the Society and hoped to be elected as a fellow. This may have been a stimulus for his later adventures.

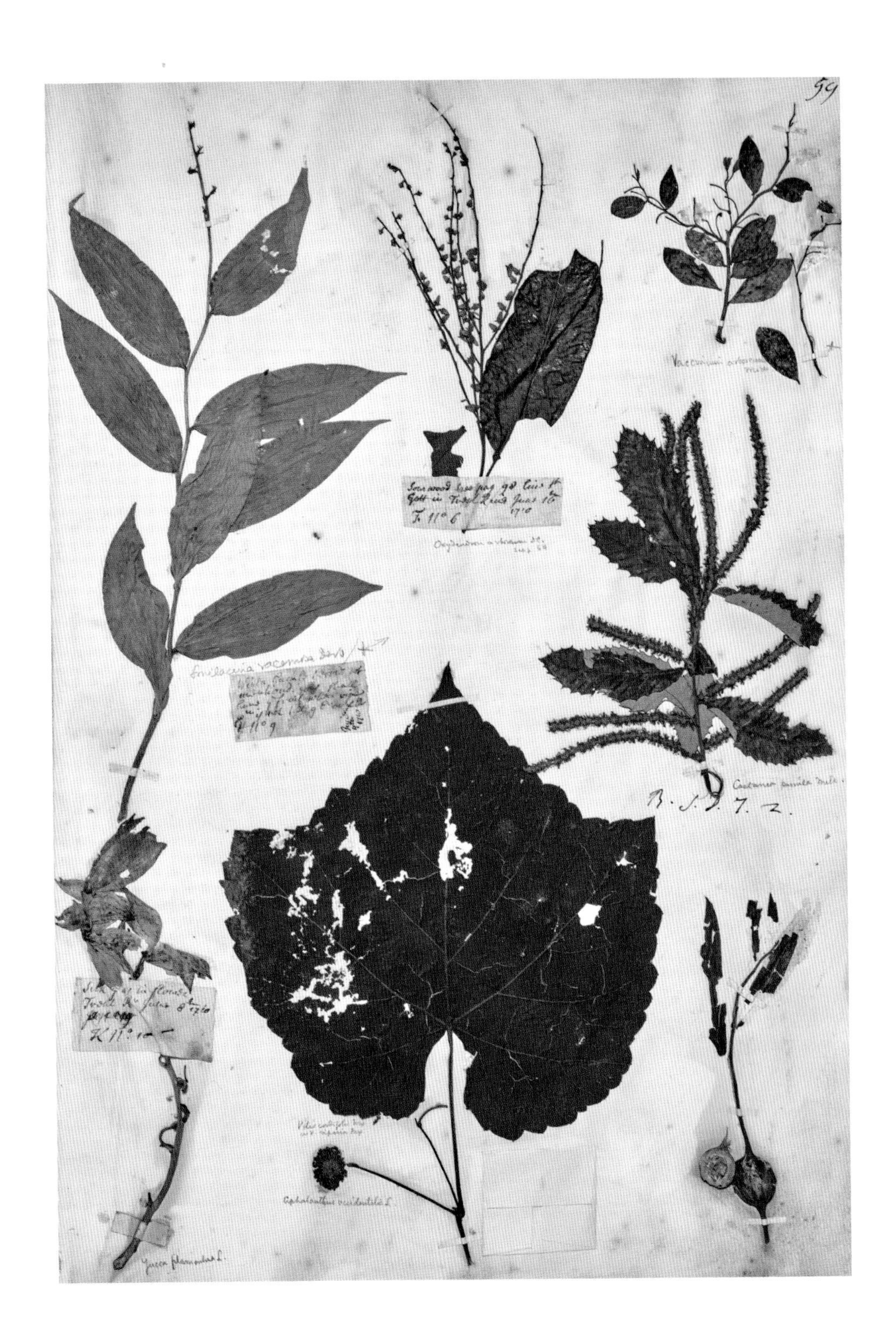

The year 1700 was the Jubilee or Holy Year celebrations of the Catholic Church in Rome and like many young men of his social standing, the 26-year-old Lawson planned to travel there. Had he followed this path, he would not have been diverted by a fortuitous encounter that he writes of in *A New Voyage*:

> *I accidentally met with a Gentleman, who had been Abroad, and was very well acquainted with the Ways of Living in both Indies; of whom, having made Enquiry concerning them, he assur'd me, that Carolina was the best Country I could go to and, that there then lay a Ship in the Thames, in which I might have my Passage.*

Lawson does not name the mysterious informant but there is evidence to suggest that it was his friend James Moore, a politician from the Carolina capital Charles Town (now Charleston), who had conducted several expeditions into the colony's interior and who would have been in London at the time. Moore secured Lawson's passage to Carolina on his ship and later assisted him when he arrived in Charles Town.

Lawson sailed from London for the Americas, arriving in Charles Town in September 1700. He found the town and its inhabitants (many of whom had made their fortunes in the surrounding plantations) to be pleasant. He soon found employment in Carolina as a freelance surveyor, subsequently becoming deputy surveyor of the colony in 1705. He was eventually appointed surveyor general in 1708 by the Lords Proprietors, the London-based governing body of the Carolina colony.

Once established in Carolina, Lawson became acquainted with local plant collectors Edmund Bohun, Robert Ellis and Major William Halsteed, who were all correspondents with the London apothecary and collector James Petiver. Lawson may have known Petiver before he travelled to the colonies as he had spent some time as an apprentice in the London Society of Apothecaries and would surely have known him at least by reputation. Lawson, perhaps with an eye on Royal Society fellowship wrote to Petiver offering to collect animals and plants for him 'I shall be very industrious in that Employ I hope to yr satisfaction & my own, thinking it a more than sufficient Reward to have ye Conversation of so great a Vertuosi'. There is no evidence that Petiver responded and the relationship would only really become established later, in 1709, when they met in person.

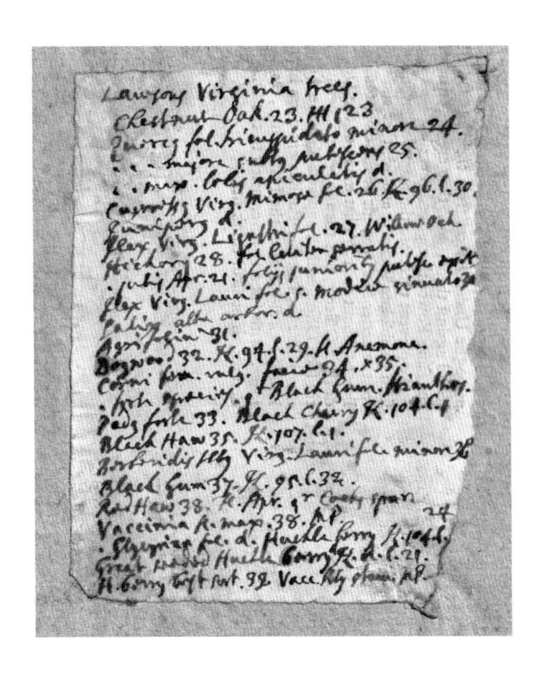
Lawsons Virginia trees.
Chestnut Oak. 23.

Above List of John Lawson's Virginia trees represented in the Sloane Herbarium in Petiver's hand. The first on the list is the 'Chestnut Oak' that Lawson described in his *A New Voyage*.

Lawson's development as an explorer and collector began soon after his arrival in Carolina. On 23 December 1700, with a group of five European companions and four native American guides, he headed into the back lands of what are now North and South Carolina. Taking a horseshoe-shaped route and beginning with an exploration of islands in the Swanee River by canoe, they then took to the land on a trek that was to last 57 days and cover an impressive 885 km (550 miles). Lawson collected and made extensive notes on his observations of the fauna, flora and the native inhabitants that they encountered including descriptions of the depredations of alcohol and small pox introduced to their society by the European settlers.

Lawson compiled his observations into a manuscript and sailed to London in 1709 to begin the process of publishing it. The first instalment was released in April 1709 with the publisher expecting further volumes after Lawson returned to Carolina. Whilst in London, he was to finally meet Petiver who had read and been very impressed with *A New Voyage* in particular his descriptions of trees that he entertainingly describes, noting their uses. For example, he writes on the swamp chestnut oak (*Quercus michauxii*):

> *Chestnut-Oak, is a very lofty tree, clear of Boughs and Limbs for fifty or 60 foot. They bear sometimes through four or five foot through all clear Timber; and are the largest Oaks we have, yielding the fairest Plank. They grow*

chiefly in low Land, that is stiff and rich. I have seen of them so high, that a good Gun could not reach a Turkey, tho' loaded with Swan-Shot: They are call'd Chesnut, because of the Largeness and Sweetness of the Acorns.

Petiver furnished him with instructions and supplies for collecting including the sound if somewhat obvious advice, 'a few pins yt[yet?] your Insects may not fly away after you have once caught them'. As an aid to his botanical collecting Petiver also provided Lawson with a copy of John Ray's account of Virginia plants collected by the Anglican minster and naturalist John Banister in 1680.

Lawson eventually sailed for Norfolk, Virginia, in January 1710, taking with him a party of Swiss and German protestant emigrants for whom he had arranged passage with the Swiss nobleman Baron Christoph von Graffenried who had purchased land (on Lawson's recommendation) to create a new settlement in Carolina. One of the two ships was detained by a French privateer operating out of Martinique and was delayed for some 13 weeks behind Lawson's vessel. Lawson used his time in Norfolk to good advantage and the first of his specimens in the Sloane Herbarium bears a note stating that it was 'gotten in Norfolk County, Virginia'. It was *Vaccinium stamineum* or as Lawson described it '...the largest huckleberry...'

All of the 295 plant specimens collected by Lawson in the Americas and now in the Sloane Herbarium were collected on his return to the colonies in 1710. There may have been specimens collected and sent to Europe before then but they are as yet unknown. There are about 100 plant species and 47

Below Specimens collected by John Lawson at Norfolk, Virginia on route to the Carolinas. The squaw-huckleberry (*Vaccinium stamineum*), (top left), is described as 'the largest Huckleberry'.

families represented, including many trees but also herbaceous plants and perhaps surprisingly only one non-vascular plant, a moss. Lawson's specimens are of good quality and Petiver and other later botanists have attached tags in Lawson's handwriting recording collection details and in many, a reference to a page in *A New Voyage*.

The specimen of a European plant, Curly Dock (*Rumex crispus*) introduced by settlers and now a common weed in the Carolinas emphasizes the value of these collections in studying changes in the biodiversity of the region.

Lawson's entrepreneurial spirit had provided him with a good living and a senior position in the colonial administration. However, his land dealing and development activities at a time of high tension and violent encounters between settlers and the local Tuscarora people were to lead to his early death. Von Graffenried planned to cross wild and European grape varieties and establish a vineyard and with this in mind set off in 1711 with Lawson, and an enslaved African to carry their baggage, to lands around the Neuse River. The three were captured by the Tuscarora and put on trial, with some justification, for crimes against the tribe. Von Graffenried was released, possibly as his fine clothes suggested he might be the Governor of Carolina, but Lawson is reported to have been tortured to death in gruesome fashion: 'they stuck him full of fine small splinters of torchwood, like hoggs' bristles, and so set them gradually on fire.'

In his short life John Lawson made one of the greatest contributions to our knowledge of not only the fauna and flora but also the language and customs of the native peoples of the Carolina colony. The *New Voyage to Carolina* is a wonderful travelogue that makes entertaining reading to this day. His name is perpetuated in the name of the town of Lawson's Creek, Carolina on the site where Lawson had built his house.

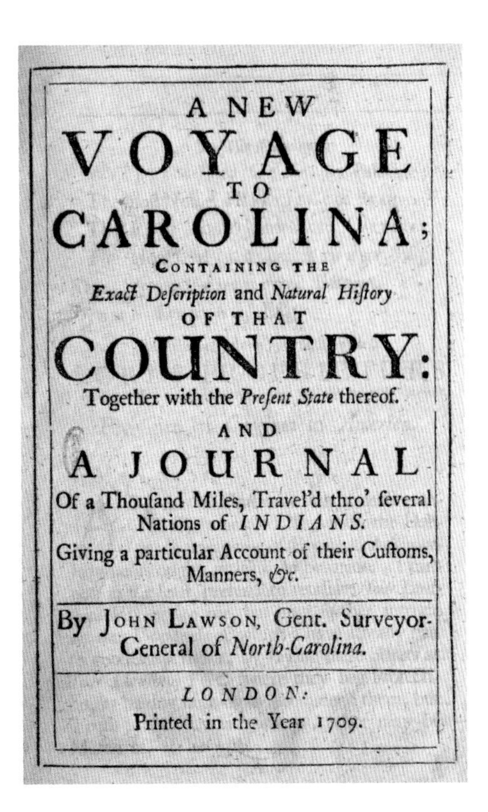
A NEW
VOYAGE
TO
CAROLINA;
CONTAINING THE
Exact Description and *Natural History*
OF THAT
COUNTRY:
Together with the *Present State* thereof.
AND
A JOURNAL
Of a Thousand Miles, Travel'd thro' several Nations of *INDIANS*.
Giving a particular Account of their Customs, Manners, *&c.*

By JOHN LAWSON, Gent. Surveyor-General of *North-Carolina*.

LONDON:
Printed in the Year 1709.

Above Frontispiece from John Lawson's *A New Voyage to Carolina*.

WILLIAM HOUSTOUN

EDWIN ROSE

William Houstoun (*c.* 1695–1715) was a Scottish physician and botanist who had originally studied at the University of St Andrews, although he left before he completed a degree. In around 1727, Houstoun became acquainted with Sir Hans Sloane and Philip Miller, the head gardener at Chelsea Physic Garden, with whom he corresponded and exchanged natural history specimens. Many of these are now held in the Sloane Herbarium.

Opposite A specimen of the epiphyte *Tillandsia polystachia* collected by Houstoun in Veracruz.

After meeting with Sloane and Miller, Houstoun embarked on a journey to Europe through France and the Low Countries before arriving at Leiden, where he studied for a medical degree under the tutelage of the eminent botanist and physician Herman Boerhaave, and graduated in around 1729. During his time in Europe, Houstoun acted as an agent for Sloane, visiting many of the European botanical gardens while compiling catalogues of their collections and obtaining specimens. For example, in early 1730 Houstoun compiled a '*Catalogues plantarum in horto Regio*' of the Jardin du Roi in Paris, which he then forwarded to Sloane.

Houstoun returned to London shortly after his degree, following which Sloane appointed him as one of the first paid collectors in the Americas. Financed by the Trustees of the Provence of Georgia, Houstoun's chief responsibility was to collect plants for a new botanical garden in Savannah. The majority of specimens from Houstoun in the Sloane Herbarium were collected on this journey and many were sent back to Sloane from Jamaica and Veracruz (Mexico). Those Sloane and his curators stored in HS 292 are labelled as 'Plants collected from Jamaica Vera Cruz and Cuba by Mr. Houstoun' and several have been labelled with Houston's original labels.

By December 1730 Houstoun had reached Kingston, Jamaica, sending a letter to Sloane alongside 'a Collection of Plants and

14

other natural Curiosities'. In addition to plant specimens, Houstoun sent several descriptions and a volume of illustrations to Sloane's neighbour, Philip Miller. A typical example is the potbelly airplant (*Tillandsia paucifolia*), an epiphyte Houstoun sent back from Jamaica alongside an accompanying illustration. Other epiphytes Houstoun collected include a large specimen of another species of Tillandsia, *Tillandsia polystachia*, that would have been challenging to collect and preserve, owing to the significant amount of effort to ensure that the plant itself was dead before he wrapped it in paper to be sent to Sloane (see image on previous page).

Above The epiphyte (*Tillandsia paucifolia*) Houstoun sent to Sloane and Miller.

In a report sent to Sloane, Houstoun commented on the 'great many Plants on the Continent [South American mainland] which I could not possibly reduce to any Genus yet described, and therefore have made bold to characterize some of them'. In part, this was a consequence of Houstoun being unable to carry a significant number of botanical works in the field for the purposes of identification but he did collect a significant number of new species that were later identified in Europe. The illustration of plants in their living state was essential for allowing botanists in London to ascribe names to those new species.

On 5 March 1731 Houstoun wrote to Sloane from Veracruz describing how the ship he had been travelling on 'was unfortunately drove ashore here and lost the 6th of last month'. However, Veracruz proved to be a particularly good botanical collecting site for Houstoun; material from this locality is scattered throughout HS 292 and 316. Houstoun was unable to obtain permission from the Spanish governor to explore the province of Jalapa and he described to Sloane how he 'sent up an Indian who has brought me down 4 small roots of it [Jalapa, *Mirabilis jalapa*] which I hope will grow and I believe we shall find it a plant quite different from the marvel of Peru'. Jalapa is a plant that has both medicinal and ornamental properties. The medical applications of Jalapa as a purgative would have been of interest to Houstoun and Sloane, whose desires to find new remedies to aid their medical practices was a major motivation for their botanical collecting. Other species Houstoun collected in Veracruz included a specimen of the water lettuce (*Pista stratiodes*), which he illustrated due to the difficulties encountered when attempting to preserve this aquatic plant. Many of these illustrations were compiled into a volume and sent to Miller in 1731.

Several of the specimens Houstoun sent to Sloane were inserted between the pages of a printed book. These were glued onto the printed pages of this publication before they were cut up by Sloane's curators and inserted into the larger volumes of the herbarium. Rather than an English, Spanish or South American book, it is apparent from the distinctive title page that the book from which these fragments came is a copy of Bonifacio Vitalini's *Super maleficiis, opus egregium totum iuris antiquorum prudentiam continens..., cum additionibus perutiliss. do. Benedicti Vadi... aliorumque clarissi. Doctorum*, a law book published in Venice by Philippum Pincium in 1518. It is probable this fairly antiquated and out of date legal tome was being sold cheaply and thus provided a source of paper that Houstoun could use to dry and preserve plants.

Below Houstoun's illustration of the water lettuce (*Pista stratiodes*) which he collected in Veracruz and illustrated to preserve vestiges of its physical shape lost during the drying process. These were compiled to form a volume Houstoun sent to Miller in 1731.

Many of the specimens attached to these printed pages have been laid out and labelled by Houstoun. However, some specimens have a firmly European origin. They include plants that Houstoun collected on his travels around Europe and those introduced to the Americas from the mid-sixteenth century, examples being the Mediterranean species of the poppy anomie (or Spanish marigold, *Anemone coronaria*), several different flowers from which Houstoun laid out on a page. Others include species that were originally from central Africa and would have been brought to the

Left A selection of different flowers from the Spanish marigold (*Anemone coronaria*) Houstoun laid out on a printed page from Vitalini's book.

West Indies and Central America through the slave trade from the sixteenth century. A typical example is a specimen of the lesser meadow-rue (*Thalictrum minus*).

In his final letter to Sloane, Houstoun offered his 'humble service to Dr. Mortimer & Dr. Amman'. Cromwell Mortimer, a secretary of the Royal Society, and Johann Amman, a Swiss botanist, were curators responsible for Sloane's collection during the 1730s. Upon receipt of Houstoun's material, it was their responsibility to insert it into the herbarium, along with adding reference codes or the polynomial names for new species to Sloane's annotated copies of John Ray's *Historia Plantarum* and Sloane's own great work, *A Voyage to Jamaica*. Mortimer's hand appears on several of the labels used for Houstoun's specimens, adding the note 'Houst. From Jamaica' after the polynomial name. New polynomials were then added to the extensive margins of Sloane's own copy of *A Voyage to Jamaica*. This was a typical feature of Sloane's Baconian approach to adding additional information to a publication, even if a new edition was never intended, treating it as always unfinished and capable of serving as a repository that could incorporate and order new information on the botany of the West Indies. Similarly, Amman was responsible for adding reference codes and descriptions of species sent by Houstoun to the extended margins of Sloane's copy of Ray's *Historia Plantarum*, allowing users of the herbarium to cross reference these physical objects with the publications.

This was the last major batch of specimens Houstoun sent to Sloane. On 14 August 1733 Houstoun died in Kingston, Jamaica, just over a year after his election as a Fellow of the Royal Society. In spite of his early death, Houstoun remains one of the earliest plant

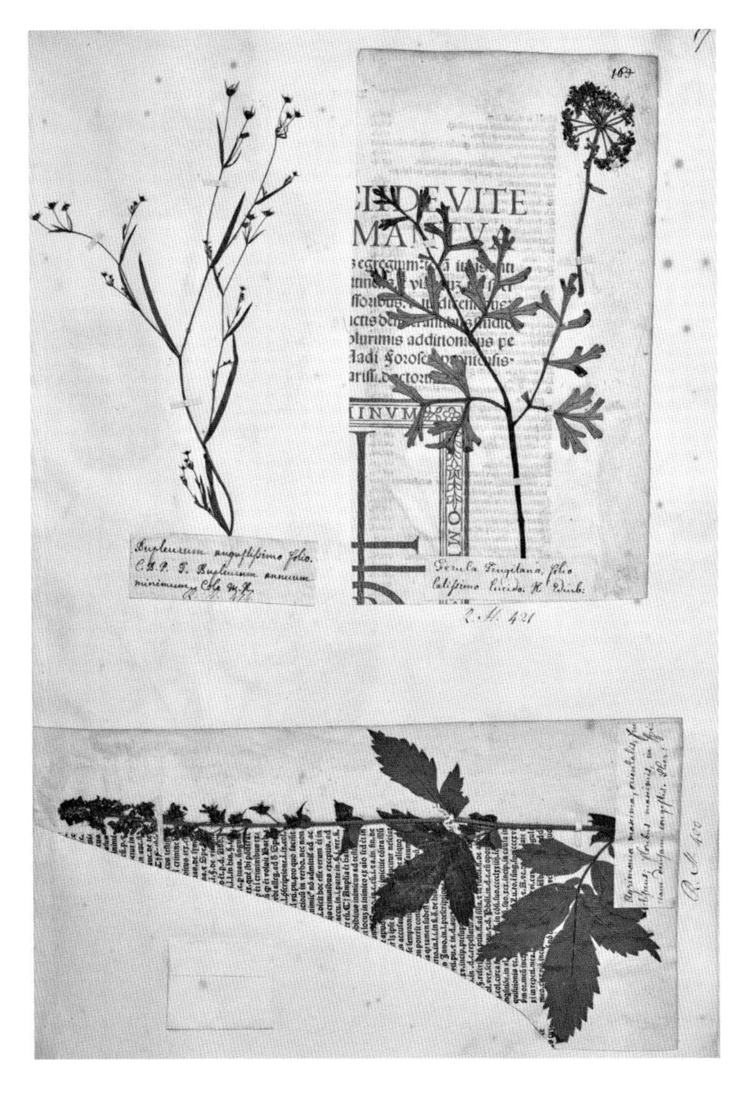

Left Some surviving fragments of Bonifacio Vitalini's Venetian legal work published in 1518 which Houstoun used to dry and mount botanical specimens from the West Indies.

collectors specifically employed to search for and collect new species in the Americas. The new species Houstoun discovered in the West Indies were valued by Miller and Sloane, and helped to establish his reputation long after his death. In 1753 the Swedish naturalist Carl Linnaeus named the genus *Houstonia*, in Houstoun's honour, sealing his name in the new binomial botanical nomenclature.

DAVID KRIEG

ROBERT HUXLEY

This curious Gentleman, after he had made several Remarks on the Natural Productions of this Island, and Painted several things he had here observed; was pleased to make a Voyage to Maryland, from whence he returned plentifully stored with what in Nature he had there taken notice of. His happy Genius in Designing, Painting, Etching, etc. were no small Additions to his other Qualifications; he being no less versed in the Study of Physick, than Anatomy, Botany, Chimistry, Natural Philosophy, and indeed whatever else is requisite to compleat a Physician.

Published in his *Musei Petiveriani* by the distinguished apothecary James Petiver, this account acknowledged the abilities of his friend, the German physician and natural historian David Krieg. Through his collecting activities in the fledgling Maryland colony in North America, Krieg is recognized as a major contributor to the development of scientific knowledge of the fauna and flora of North America.

David Krieg spent most of his adult life as a practicing surgeon in Riga, Livonia (now Latvia) but the majority of his significant achievements in natural history were accomplished within a short window of time between 1696 when he moved to London, and 1698 when he returned from a five-month trip to America.

He was born around 1669 in Annaberg, a small town in the mountains of lower Saxony close to what is now the border of Germany and the Czech Republic, and he was admitted as a surgery student in 1691 at the University of Leipzig. It was there that his interests in botany began to take form. Up to the late nineteenth century, it was common practice for botany to be taught as an integral part of courses in surgery and medicine and at Leipzig,

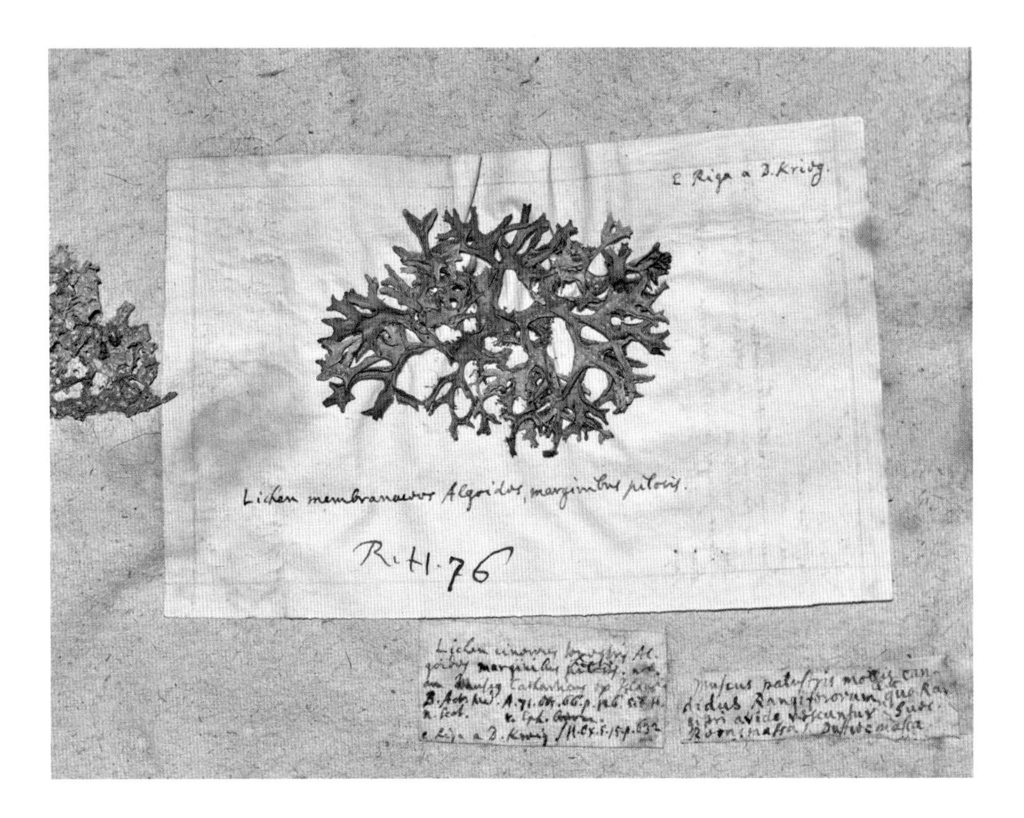

Above Lichen collected in the region of Riga, by David Krieg and sent to James Petiver in London.

Krieg would have participated in botanical lectures and field expeditions to collect plants of medicinal value. Much of Krieg's botanical training was under the able tutelage of Augustus Quirinus Rivinus, a pioneer in plant classification, who corresponded and debated with the great John Ray on this topic. Krieg would later meet Ray during his travels in England.

From Leipzig, Krieg returned to his home in Saxony for a brief spell arriving in Riga in 1694. Riga was not a random choice as Krieg's brother, Elias, had worked there since 1674. Despite the presence of his brother in the city, Krieg did not settle and after a short spell travelled to Utrecht to defend his doctoral thesis and then to London, arriving in 1697.

The University of Leipzig had not only produced in Krieg a skilled surgeon but also an accomplished natural historian and illustrator of specimens and in London it was these talents that brought Krieg to the attention of the English natural history community. He attended the Friday meetings of the botanical club at the Temple Coffee House where he expanded his natural history knowledge in the company of Sir Hans Sloane and his 'curious friends'. Petiver offered the able Saxon lodgings in his house and soon they became close friends. He also put Krieg to work drawing sketches of specimens from his collections. Through his travels in England, Krieg was to meet other notable natural historians of the day: at Oxford he met Jacob Bobart, Professor of the Physick Garden, and in Essex he was introduced to John Ray, a correspondent with Rivinus, his former botany professor.

Krieg's presence in England coincided with a great demand from the botanical community for plant specimens from temperate America and it was a natural progression in his development when Petiver encouraged him to travel to the American colonies to collect specimens. Krieg departed for the Maryland colony in the ship *John and Thomas* in January 1698, earning his passage as ship's surgeon. William Vernon, the Cambridge-educated naturalist and fellow member of the botanical club at the Temple Coffee House, had departed for Maryland some weeks earlier. Vernon was supported by William Nicholson, the governor of the colony but Krieg had to earn a living putting his medical training to good use in treating the Maryland planters in exchange for tobacco, the currency of the colony at that time. During his relatively short stay he also worked as a forensic pathologist. The founder of this discipline was teaching at Leipzig when Krieg was a student but it is not known if their paths crossed.

Krieg arrived in Maryland in April 1698, a month after Vernon. His first month did not go as well as expected and he only collected 22 plants and a few insects. This he partly blamed on the unavailability of his 'assistant' Isaac, or 'Sancho' as he was known, who had been employed by Petiver to help Krieg and Vernon. Isaac also served as cabin boy for the captain of the *John and Thomas*, which may explain the limited assistance he provided.

Despite his disappointing first month, a particular strength of Krieg's Maryland collections was the number of spring-flowering plants that he was able to gather and send back to England. Though he was in Maryland at the same time as Vernon, their movements remained separate. Furthermore, Vernon was expecting to be in the colonies for several years and did not start collecting in earnest on his arrival. It was therefore particularly fortunate for the London botanists that Krieg had so eagerly collected spring flowers since Vernon's trip was curtailed and he returned to England later that year following the recall of his patron, the governor, without further opportunities for collecting.

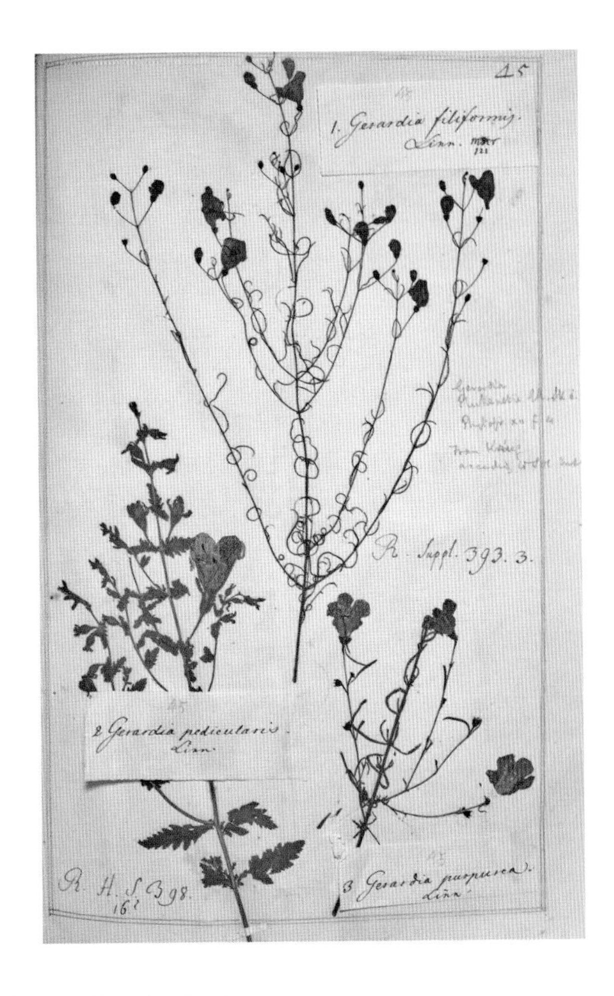

Above Slender false foxglove (now named *Agalinis tenuifolia*), collected by David Krieg in Maryland (specimen at top of page). The slips of paper with plant names were added later by Daniel Solander.

Krieg spent five productive months in Maryland. He sailed for London in September 1698 arriving in November that year to some acclaim from those who had put in orders for Maryland specimens. He was subsequently rewarded for his efforts by election to the Royal Society.

While Krieg collected a wide range of Maryland plants, the usefulness of his specimens was found to be lacking. Krieg had not labelled specimens with the vital details of date and place of collection and moreover many specimens lacked flowers and other parts essential to their identification. Ray, who had received the Maryland plants of both Krieg and Vernon to identify and classify, complained to Sloane: 'Those that gathered them might easily have been given an account of these, as also of the place they were found... I am at a loss what to do in this case, and want your advice. I was in hopes that both the collectors and yourself had named them, added some notes and remarks upon them; but, alas! I find none,

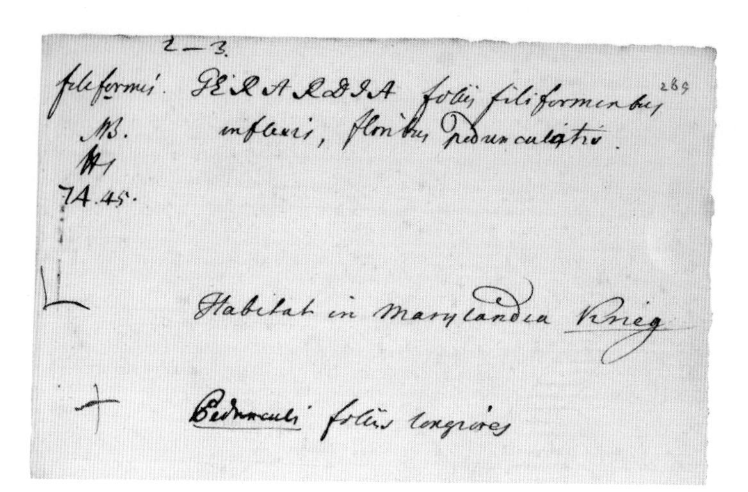
2—3.

filiformis. GERARDIA foliis filiformibus, inflexis, floribus pedunculatis.

MB.

74.45.

Habitat in Marylandia Krieg

Pedunculi foliis longiores

Left Page from Daniel Solander's slip catalogue referring to the specimen of '*Gerardia filiformis*' (now *Agalinis tenuifolia*) in the Sloane Herbarium. The locality and collector are recorded as 'Habitat in Marylandia Krieg'.

so that I am in a wilderness, and at a great uncertainty.' This lack of information remains a barrier to students of Krieg's specimens and it can be difficult to distinguish between collections made by Krieg and by fellow Maryland collector, William Vernon. In the case of Krieg's specimen of the slender false foxglove (now named *Agalinis tenuifolia*), the only clues to the collector are found in the 'slip catalogue' compiled for the British Museum in the 1760s by eminent Swedish botanist Daniel Solander who worked extensively on the Krieg and Vernon collections. His slips of paper record identifications of the specimens with descriptions and references to botanical works and sometimes the name of the collector and locality, in this case Maryland and David Krieg.

Krieg left London and returned to Riga in 1699. For a Saxon like Krieg, working in a pro-Swedish city was fraught with dangers as Riga was caught up in the Great Northern War that was raging at the time. Fortunately, his origins do not seem to have affected his ability to practice and to live in the city and he continued to correspond with his English natural historian friends. Books such as Ray's *Historia Plantarum* and Sloane's *Voyage to Jamaica* arrived in Riga and in return Krieg sent books, items of Livonian culture, rock samples and insect and plant specimens. There are several specimens today in the Sloane Herbarium collected by Krieg from this period.

Like many students of the natural world at this time, Krieg did not confine his curiosity to plants and animals; he actively studied chemistry and that of glassmaking in particular, publishing a paper in the *Philosophical Transactions of the Royal Society* on the chemistry of colouring in ceramics.

The end of the Saxon siege of Riga in 1700 marked a new chapter in Krieg's botanical career with the arrival in the city of French diplomat Louis Comte de Guiscard-Magny who needed a personal physician to accompany him on his travels. Krieg, with his experience as a ship's surgeon, was ideal for the job and his work for the Comte took him to many destinations in Europe.

Krieg left the Comte's service in 1702 and returned to Riga. By 1710 the city was besieged again, on this occasion by the Russian army. Whilst the siege and his surgeon's work limited Krieg's time to collect and study natural history he did continue to botanize and correspond. He wrote to his London friends of a return to England or of settling in the West Indies where the climate would have better suited his health. These dreams were not to be realized as the plague that followed the siege was to claim his life. He died in 1710 but the effects of the war on communication meant that James Petiver did not hear of his good friend's death for three years.

Krieg's collections from coastal Maryland form an important record of the region to this day providing a baseline to highlight the dramatic changes in the vegetation over the last 300 years. His name is immortalized in the Latin name of the North American dwarf dandelion genus *Krigia*.

MARK CATESBY

STEPHEN HARRIS

Mark Catesby, one of Philip Miller's so-called 'Procurers of Plants', was a naturalist, sponsored by a group of gentlemen to explore the American colonies and the Bahamas in the early eighteenth century. His authorship of *The Natural History of Carolina, Florida and the Bahama Islands* (1729–1747) made his name among the virtuosi of Georgian England; the original illustrations were eventually acquired by George III. The living plants he helped introduce to Britain became part of the gardens of the great and the good; some even became commonplace horticultural staples. However, knowledge of his plant specimens, rapidly incorporated into the herbaria of Sir Hans Sloane, William Sherard and Charles Dubois, gradually disappeared from the sight of natural historians during the eighteenth century, only to be resurrected in the twentieth century.

Opposite The lily thorn (*Catesbaea spinosa*) and zebra swallowtail (*Protographium marcellus*) from Catesby's *The Natural History*. The lily thorn's scientific name was coined by Jan Frederik Gronovius to commemorate Catesby.

Catesby was born in Castle Hedingham, Essex, to affluent parents. He spent his early years on the Essex-Suffolk border, where his botanical interests were cultivated by his uncle Nicholas Jekyll, an acquaintance of English naturalist John Ray and the apothecary Samuel Dale. Between 1712 and 1719, Catesby lived with relatives in Virginia where he collected 'some dried specimens of plants and some of the most specious of them in tubs of earth, at the request of some curious friends'. During this period, Catesby proved his worth as a collector of American plants to natural historians in England.

When Catesby, 'who designs and paints in water-colours to perfection', returned to England, Dale was instrumental in ensuring his skills came to the attention of the influential naturalist brothers William and James Sherard. James Sherard, who owned a magnificent garden at Eltham, would eventually grow many North American species collected by Catesby.

By 1722 William Sherard was part of a group of high-profile natural historians and horticulturalists supporting Catesby's more

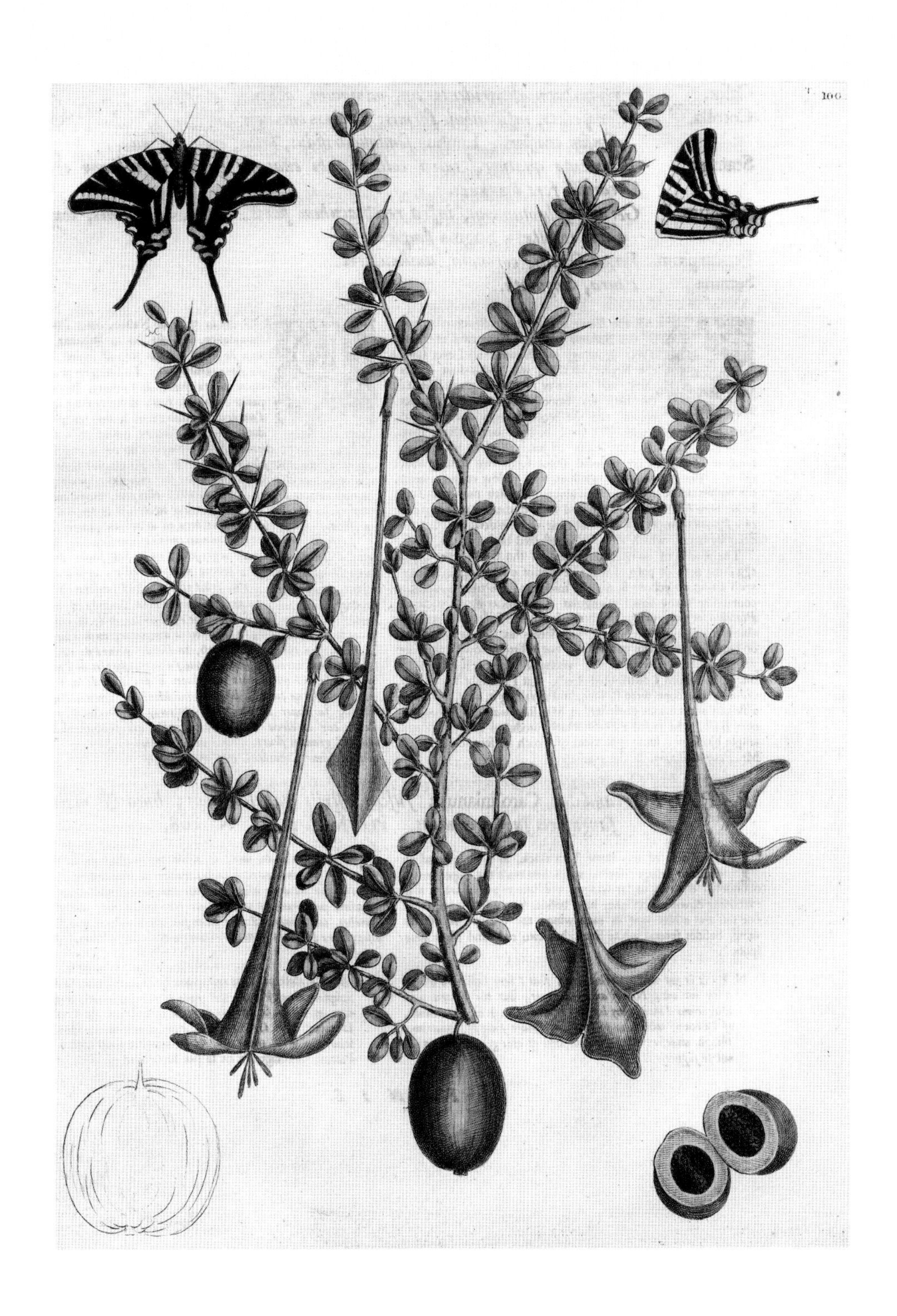

Left Catesby's sketch of the American lotus (*Nelumbo lutea*) sent to William Sherard from Colonial America. It was accompanied by pressed leaves and a note about the plant's identity and ecology.

ambitious, and focused, explorations of the American colonies. The group included Charles Dubois, the treasurer of the East India Company, the physician Richard Mead and the politician James Brydges. Horticulturalists sponsoring Catesby included the Hoxton-based nurseryman Thomas Fairchild, who went on to propagate and sell much of the living material Catesby sent back to England, and the cloth-merchant Peter Collinson. Reflecting on his expedition years after his return, Catesby reserved special praise for 'that great Naturalist and Promoter of Science Sir Hans Sloane, Bart to whose goodness I attribute much of the success I had in this undertaking.'

During the four years Catesby was sponsored to explore North America, he travelled through the Carolinas, into Georgia and Florida and finally the Bahamas. Catesby quickly realized that 'to collect everything is impossible but with many years application', even though he had the advantage of living for protracted periods in the areas he explored. His approach to collecting was 'never to be twice at one place in the same season', a strategy perhaps driven by expediency as much as by biology, since his sponsors were interested in novelties. However, most of the hundreds of extant herbarium specimens Catesby collected lack the minimum details one would expect to see on modern equivalents, such as the collection locality and date. Catesby found collecting seed and botanical specimens in the field hard work, as it involved travelling long distances, sometimes in inclement weather, over hostile terrain. Once samples were collected, the work was not over; they had to be labelled, preserved, packed and transported to England. Added to this, Catesby was also making drawings in the field. As he started to deliver on his promises, the work became harder. More sponsors wanted to support him, so he had to haul around more baggage, which would have included specimens in various stages of preparation, paper for drying plants, artist's supplies, not to mention basic food and personal supplies. Some specimens proved a challenge to preserve, even for a determined collector such as Catesby. For example, associated with a leaf specimen of the floating, aquatic American lotus (*Nelumbo lutea*) is a sketch of its flower that Catesby made in the field, together with a note to William Sherard: 'The flowr I could not preserve so have sent this scetch.'

There is little doubt therefore that he used all resources available to him to collect, including native American guides and porters, and willing friends and acquaintances. He even purchased a slave, after petitioning William Sherard for the funds: 'the next Negro Ship that arrives here I designe to buy a Negro Boy which I cannot be without.'

Catesby discovered that patronage comes with the burden of trying to satisfy the competing demands of sponsors, a particularly delicate task when one of your sponsors was such a powerful figure as Sloane, whose favour Catesby was concerned he would enjoy when

Below Catesby's specimen of Indian bean-tree (*Catalpa bignonioides*) together with his note recording visits to similar sites at different seasons. A similar label is found with a duplicate specimen in the Sloane Herbarium.

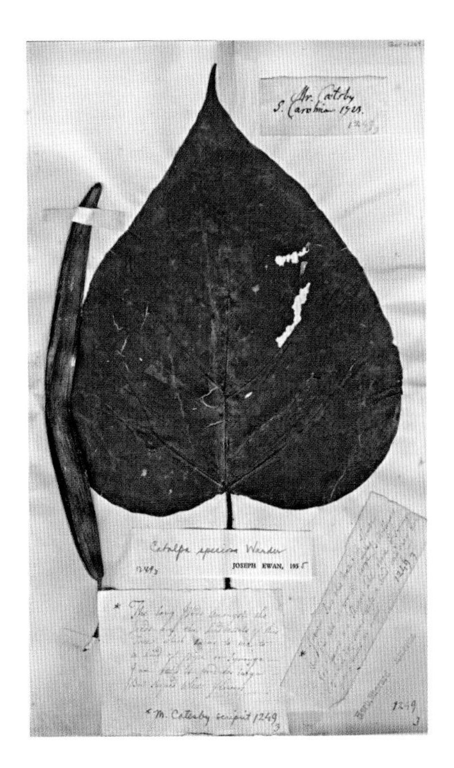

he returned to England. Dubois also proved difficult to satisfy: 'the discontent of Mr Du-Bois and the trouble he gives my Friends in receiving his Subscription is Such that I had rather be without it, I doubt Not that I have Suffered by his Complaints.'

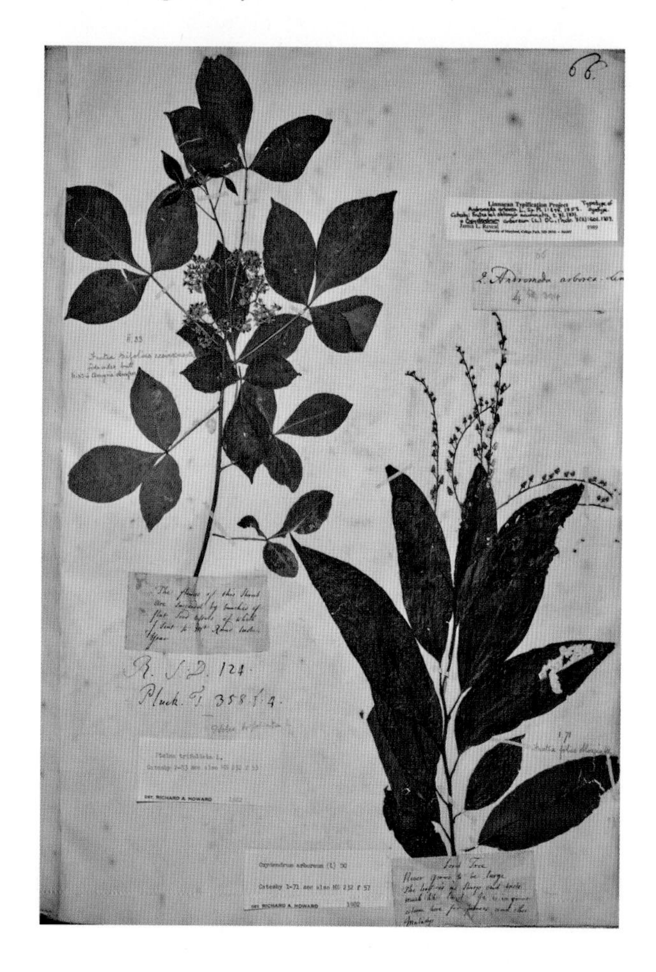

Catesby returned to England in 1726, having sent thousands of herbarium specimens and living plants, sometimes packed in precise and ingenious manners, to his English sponsors. Catesby became an advocate of growing North American trees in Britain in order to discover their economic value. He believed this could not be achieved in American colonies: 'unless by their [trees] becoming free denisons of our woods and gardens, their plenty may afford opportunities of discovering their uses and virtues; which in an infant country, little inclined to improvements, and depending on its mother country for all kinds of utensils, cannot be expected.'

In London Catesby began the protracted process of preparing illustrations and text for *The Natural History…*, making extensive use of Sloane's collections. The first 20 hand-coloured plates were issued in 1729, to very positive reviews, which set the standard for the other 200 plates. Catesby completed the publication in 1747, two years before he died. He is buried at St Luke's Church, Old Street, London.

Above Specimens of hoptree (*Ptelea trifoliata*) and sorrel tree (*Oxydendrum arboreum*) from the Sloane Herbarium, collected and annotated by Catesby.

The skill and decisions made by field collectors determine the future scientific potential of a specimen. Judgements made by collection owners determine whether specimens survive, are accessible to researchers and hence whether they become part of the corpus of scientific knowledge. In the case of Catesby's herbarium specimens, those in Sloane's herbarium have been widely used for research, whilst those in the herbaria of Sherard and Dubois have been largely overlooked. However, when the German-American botanist Friedrich Traugott Pursh was working on his early nineteenth century North American flora, he was delighted to find Catesby's specimens in William Sherard's herbarium. Surprisingly, he makes no mention of those in the Sloane Herbarium.

Below The hoptree (*Ptelea trifoliata*), a familiar garden plant introduced to Britain by Catesby, and the eastern tiger swallowtail (*Papilio glaucus*) from Catesby's *The Natural History*.

Catesby, through contacts and his skills as a collector and illustrator, successfully penetrated the strong social and intellectual networks connecting eighteenth-century naturalists, becoming a Fellow of the Royal Society in 1733. Catesby's *Natural History* remains a vivid snapshot of the activities of a naturalist-collector, and the flora and fauna of colonial America. Advances in microscopy, chemistry and computing have given us new ways of looking at, and extracting information from, the specimens Catesby collected. In Catesby's words, 'by the concurrent endeavours of the philosopher and artisan, I question not but many of them will be found useful to purposes, of which at present we have not the least conception'.

JOHN BARTRAM

VICTORIA PICKERING AND MARK CARINE

John Bartram, who has been described as the father of American botany, was a Quaker, a farmer and a plant collector, born on a farm in Marple, Pennsylvania, in 1699. Through his friendship with fellow Quaker Peter Collinson, a Fellow of the Royal Society whose gardens in London were renowned for their array of exotic species, Bartram was responsible for introducing more than 300 North American plant species into England.

The two men appear to have met through the Collinson family's extensive trade in the cloth business with North America. They became close friends and through their partnership, Bartram successfully sent seeds and seedlings from America to Collinson in England who then distributed them to naturalists (and others) across Britain and Europe. Bartram became well known for his methods of packing seeds and plants to withstand the long arduous journey across the Atlantic, including protecting them in bottles, gourds and other containers.

The centre of the operation was Bartram's farm. In 1728 Bartram purchased land from Swedish settlers in Kingsessing on the west bank of Schuylkill River, not far from the centre of Philadelphia. There he assembled his collection of American plants that underpinned his seed and plant business. His house and garden survive today and the garden is considered to be the oldest surviving botanical garden in North America.

The Bartram and Collinson distribution of seeds and plants proved popular among influential landscape planters like the Duke of Richmond, Philip Miller at the Chelsea Physic Garden, and naturalists such as Stephen Hales and John Ellis. They were all keen

Below John Bartram depicted in an illustration from an article entitled 'Bartram and His Garden', published in *Harper's New Monthly Magazine* in 1880.

to receive boxes of American seeds to cultivate in their gardens. The contents of these boxes were often mixed and included a range of seeds and dried plant specimens packed with root-balled plants, stuffed animals, preserved insects, birds' nests and fossils. However, people often requested seeds because they were far easier to transport than live plants and could be more easily distributed across Europe. By 1753, Bartram and Collinson's boxes of seeds had become a successful enterprise and cost around five guineas each. As a result of this, there was a huge influx of American plants into Britain and this had a great impact on the landscape.

Bartram may not have had the professional or social standing of other collectors but his location within the New World meant that he had local knowledge and expertise about American natural history which was highly valued. Bartram initiated written communication with Sir Hans Sloane who would go on to encourage these exchanges. Sloane evidently recognized Bartram as a useful contact to have if he wanted to add further American material to his botanical collection.

Bartram appears to have corresponded directly with Sloane for the first time in the summer of 1741 with Collinson acting as a broker for this exchange and those that followed. Bartram wrote to Sloane:

> *My good, faithful friend Peter Collinson, in his last letter to me, that I received, acquainted me that thee desired I would send thee some petrified representations of Sea Shells. Accordingly, I have sent thee a few, which I gathered toward the northward. However, I design to send thee another collection by Captain Wright [...] when I hope to give thee a fuller demonstration that I am thy vigilant and industrious friend.*

Having sent Sloane a collection of marine molluscs in the hope that they would encourage further correspondence, Bartram must have been delighted when Sloane replied and thanked him for his generous offerings. In fact, Sloane then expressed his desire 'to have some seeds, or samples of your plants, for my collections of dried herbs, fruits, &c. I should be extremely pleased to know wherein I can be useful to you, and retaliate the obligation you have laid upon'. Some historians have argued that Bartram's modest background intensified his desire for distinction and he would have relished the opportunity to send Sloane American natural history specimens. However, Sloane's letter also suggests that he did not anticipate that these would be one-way exchanges and that he did not hesitate to encourage a relationship with Bartram, sending him a copy of his *Voyage to Jamaica* (of which Volume I was published in 1707) as well as his catalogue of Jamaican plants published in 1696.

In November of 1742, Bartram gave the impression that Sloane presented him with a gift: a silver cup with his name engraved upon it 'so that when my friends drink out of it, they may see who was my benefactor'. The engraving read 'The Gift of Sir Hans Sloane, Baronet, to his Friend, John Bartram, Anno 1742.' However, this was something that Bartram rather than Sloane had commissioned, allowing, as Sloane's biographer James Delbourgo puts it, Bartram to 'collect and display Sloane as a highly desirable object in the eyes of his American friends'. Sloane did, however, give his permission for one of his five guinea payments to be used to purchase the cup and it perhaps functioned in similar ways to the visits made by other collectors to Sloane in London, to further cement the relationship with the great collector.

The relationship with Sloane would not have harmed Bartram's social standing in Pennsylvania where he was one of the founders of the American Philosophical Society, the oldest learned society in the USA, in 1743. Many of his fellow Pennsylvanian Quakers were abolitionists and it has been suggested that Bartram may have freed his own slaves but in his Quakerism he was certainly not without controversy: he spoke out against non-resistance when there were

Above Specimens collected by Bartram. The specimen of wood geranium (*Geranium maculatum*, left) has a label in Bartram's hand, recording its uses. His specimen of American ginseng (*Panax quinquefolium*, right) has detailed notes on its habit and ecology but no information on its uses.

fears of attack by French forces in 1748 and 10 years later he was disowned by the Darby Monthly Meeting, Pennsylvania for 'heresy'; the testification stating that he 'disbelieved in the divinity of our Lord and Saviour Jesus Christ and being perfectly God as well as man'.

Three volumes of plants collected by Bartram are in the Sloane Herbarium today. They include trees, herbs, ferns, bryophytes and lichens sent to Sloane in 1742 and 1743. In a letter to Sloane dated 14 November 1742, Bartram writes of 'a quire of paper filled with dry specimens of plants, numbered'. A letter the following year describes 'two quires of specimens, gathered in their full bloom — as many as I could, but several that I found amongst the Indians, could not be found with their proper characteristics'.

Detailed notes accompany many of the specimens. The label of his wood geranium (*Geranium maculatum*) specimen reveals that it '...grows plentifully in our woods & is used by many for ye bloody flux ye root powdered & drank'. Particularly detailed habitat notes are provided for his specimen of American ginseng (*Panax quinquefolium*). The label records 'this is our gin seng A remarkable plant in its places of growth which is in shady rocky rich steep receses towards ye sources of our rivers as if designed to be hid from our use or abuse: its vertues being as little known to us as its places of growth which with its great scarcity its remarkable form exceeding thinness of its leaves long life with so little increase renders it worthy of our notice'. It is surprising that Bartram records no uses for this plant, the roots and leaves of which were used in traditional medicine by Native Americans. In China, American ginseng came to be considered superior to Asian ginseng and harvesting for export to the traditional Chinese medicine market from the mid-eighteenth century led to its marked decline.

In his second consignment of plant specimens Bartram '... also wrapped up, in separate papers, 'several of our North American Mosses'. Bryophytes were not initially a focus of Bartram's collecting, no doubt given the focus of his seed and plant business. In an undated letter to Catesby he commented 'Before Doctor Dillenius gave me a hint of it, I took no particular notice of Mosses, but looked upon them as a cow looks at a pair of new barn doors; yet now he is pleased to say, I have made a good progress in that branch of Botany, which really is a very curious part of vegetation'. The Oxford professor Johann Dillenius he referred to was the author of *Historia muscorum*, a natural history of mosses, published in 1741.

Reflecting his interests in non-flowering plants, Bartram's specimens in the Sloane Herbarium include many mosses and lichens. It also includes a plant of 'Spanish Moss' (*Tillandsia usneoides*) that Bartram took for a true moss. It is, in reality, a peculiar species of flowering plant that grows epiphytically on the branches of trees and has the appearance of a moss or lichen. Bartram records on the label 'this moss I brought several years ago from ye eastern shore of Virginia where it growth above 6 foot long and is a fine food for horses...'; Bartram travelled widely hunting

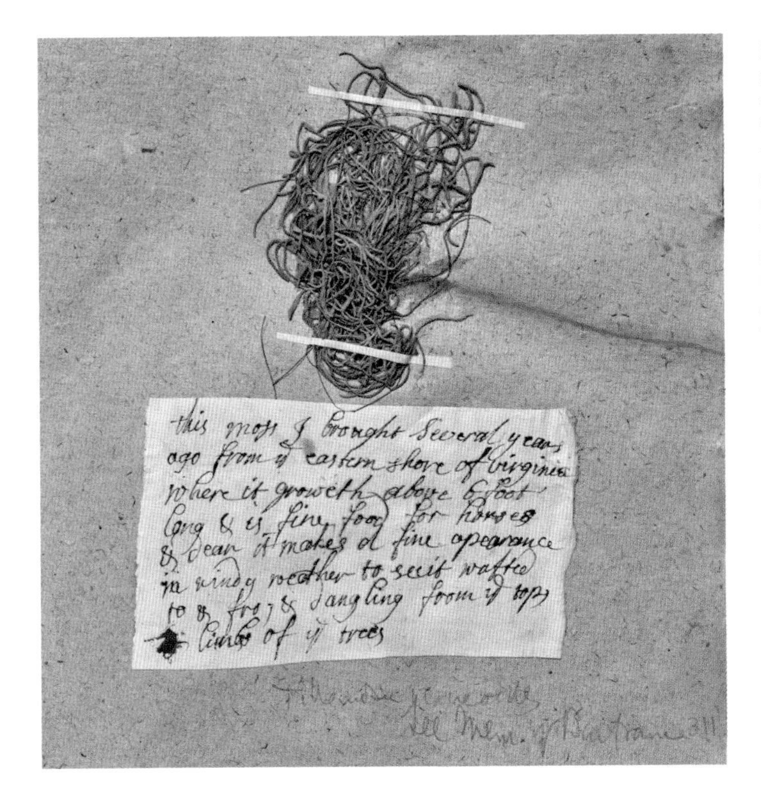

Left A specimen of Spanish moss (*Tillandsia usneoides*) collected by Bartram in Virginia, probably during his 1738 Blue Ridge Mountain trek. This is a flowering plant that grows epiphytically on trees. As Bartram records, it can reach several metres in length and 'it makes a fine appearance in windy weather to see it wafted to and fro…'

for plants as far south as Florida, north into New England and west to Lake Ontario. As his labels reveal, he was a keen observer and documenter of the plants he collected, their habitats and their uses.

In the years after Sloane's death, Bartram would become one of the first Linnaean botanists in North America. Linnaeus famously referred to him as the 'greatest natural botanist in the world'. His expertise was further recognized by his receipt of an annual pension of £50 to collect plants for the crown from 1765 onwards.

A self-taught man, John Bartram, was highly regarded by his patrons, including Sloane. Unusually among the contributors to the Sloane Herbarium, hunting for new and garden-worthy plants was his business and it was a very successful one. With his friend Peter Collinson he was responsible for introducing into cultivation in Europe hundreds of North American plant species.

COLLECTORS IN THE FIELD: ASIA AND THE PACIFIC

MARK CARINE AND ROBERT HUXLEY

Through his own observations and the collections he made in the field, Sloane gained first-hand knowledge of plants in Britain and Europe and the wider Atlantic World. For Asia and the Pacific that was not the case. He travelled to neither Asia nor the Pacific and whilst he would have seen some of the region's plants in English gardens, in most cases the dried specimens in the herbarium would have been his closest contact with the plants of this region.

Opposite Plants collected by Engelbert Kaempfer in Japan in 1691/2.

In Asia, plant collecting largely followed trade. In Western Asia merchants and officials of the Levant Company, founded in 1581 and responsible for English trade with the Ottoman Empire and the Levant, were among the contributors to the Sloane Herbarium. William Sherard, who featured in the chapter on the great collectors and cataloguers, was the Levant Company's consul in Smyrna in modern-day Turkey and sent specimens to Sloane from there. James Petiver received specimens from James Braylsford, whom he described as a 'Turkey Me[r]chant', the term used to refer to members of the Levant Company. Petiver noted that 'This worthy Gentleman was pleased freely to present me... with Four Bookes of Plants which he gathered about Jerusalem, the Mountains of Bilan and on the Banks of Euphrates and Jordan...' The Levant Company and its servants provided Sloane and other English naturalists with a window into the rich plant diversity of the eastern Mediterranean and western Asia.

Further east, employees of the East India Company (EIC) were among the contributors to the herbarium. The EIC was established to trade east of the Cape of Good Hope and west of the Straits of Magellan and the initial foci of its activities were India and

Southeast Asia. Later, fuelled by demand for Chinese goods such as tea, porcelain and silk, trading posts were also established in China. Company employees contributed plants to the Sloane Herbarium from all of those regions and, indeed, from further afield. There are, for example, collections from St Helena in the South Atlantic, an island governed by the EIC and from the Cape in South Africa where their ships called.

It was not only English trade with Asia that brought specimens to Sloane and his herbarium. The Dutch East India Company (Vereenigde Oostindische Compagnie or VOC) was the EIC's main competitor. The Dutch were the only Europeans not to be excluded from Japan in the Sakoku edict of 1635 and Sloane's Japanese specimens are the result of Dutch trade. Some Asian specimens in the Sloane Herbarium were not directly linked to trade at all. There are specimens from Siberia sent to Sloane by Johann Amman, his one-time assistant who was professor of botany at St Petersburg in Russia. There are also specimens from the Levant collected during an expedition by Joseph Pitton de Tournefort whose lectures Sloane had attended at the Jardin du Roi in Paris years earlier. Yet another source of specimens were Christian missionaries.

Here we look at four Asian collectors. The surgeon Samuel Browne was one of a number of EIC contributors to the herbarium. He was a surgeon at Fort St George in modern-day Chennai, India. Browne led what could certainly be considered an eventful life, but he also collected specimens that are of particular interest because of the way in which he worked with local doctors to record the local names and uses of the plants he collected. The Scot James Cuninghame, another surgeon associated with the EIC, made at least two trips to China. His collections are the earliest known from the country and, with his focus on useful plants, they include the earliest surviving sample of tea in Britain. The German-born VOC surgeon Engelbert Kaempfer made collections in Japan that have been considered among the most important in the entire Sloane Herbarium. Purchased by Sloane after Kaempfer's death, they are the earliest herbarium collections from Japan.

The Czech missionary Georg Joseph Kamel was also among the contributors of specimens from Asia. Kamel was a Jesuit, a member of the Society of Jesus, whose missionaries travelled widely seeking converts to Catholicism and exchanging knowledge along the way. In Asia the Jesuits established relations with China long before the EIC or VOC and their sphere of influence was wide. Kamel was both a missionary and an apothecary in the Philippines, then a Spanish territory. The essay on Kamel – after whom *Camellia* was named – explains how the plants he collected from the furthest reaches of Asia came to be in the Sloane Herbarium today.

Our final contributor is George Handisyd, who collected in the Pacific rather than Asia. A ship's surgeon, Handisyd collected plants in the perilous Straits of Magellan, at the southern tip of South America, through which ships passed to the Pacific from the Atlantic. He also collected in the Pacific, in places including the uninhabited Juan Fernandez Islands, where he collected plants nearly a decade before Alexander Selkirk – the inspiration for Robinson Crusoe – was famously marooned there. Selkirk was a privateer, engaged in activities in the Pacific of dubious legality. Handisyd, considered by Sloane to be a good friend and 'ingenious gentleman', might well have been one too.

Below A sample of *Ferula assa-foetida* from Sloane's Vegetable Substances collection. This species is one of the sources of asafoetida, widely used in Indian cooking.

SAMUEL BROWNE

ANNA WINTERBOTTOM AND RANEE PRAKASH

Samuel Browne was an East India Company surgeon stationed at Fort St George, the Company's headquarters in Madras India (now Chennai). He contributed a large number of specimens to the Sloane Herbarium. Most arrived via the collections of James Petiver and some via Leonard Plukenet. Browne also relayed specimens collected by other botanists, notably Georg Joseph Kamel, who also features in this book.

Nothing is known of Browne's early life. Perhaps, like many of the Company's surgeons, he began his career as an apprentice surgeon or apothecary, although he does not appear in the records of the Society of Apothecaries. While Browne used English in his own descriptions of plants, he was literate in Latin and his correspondence with Kamel suggests that he also read Spanish.

The earliest collection associated with Browne in the Sloane Herbarium is in a volume dated 1686. This is a collection of rare plants indexed by Browne and appended to a collection of plants made around London by Leonard Plukenet. This suggests that Browne was already botanizing, had some knowledge of exotic plants and had formed an association with Plukenet by this stage. The collection includes specimens of *Moringa oleifera*, the drumstick tree, an Asian species with both culinary and medicinal uses. It also includes a specimen described as 'Arbor Chinchina' in reference to the cinchona tree, the bark of which was used to treat malaria. However, the latter identification seems to be incorrect, demonstrating that the source of the so-called 'Jesuit's bark' remained a mystery in seventeenth-century England.

Browne first appears in the Company records as the surgeon of a ship, the *Dragon*. He was appointed surgeon at Fort St George, the English settlement in Madras, in May 1688, following the sudden death of the previous surgeon, Dr Heathfield. The same

year, he married Ann Baker, with whom he had a son and daughter. Browne's correspondence with James Petiver began in 1689, when Petiver wrote to introduce himself. The two would correspond until Browne's death in 1698. Many of their letters are preserved among the Sloane manuscripts in the British Library. Browne and Petiver had overlapping interests in medicine and botany and they corresponded on plants, insects and also touched on animals.

Above Specimens indexed by Samuel Browne including *Moringa oleifolia* (left folio, top right) and a species that he referred to as 'Arbor Chinchina' (left folio, top left) in reference to cinchona.

Madras had been established in 1639 and was the Company's most important settlement by the late seventeenth century. Its population was around 80,000 by the early eighteenth century, when Herman Moll produced his map of the city. Textile production and trade were key to the city's success, and its diverse population included Armenian and Parsi traders as well as Portuguese and African soldiers. The local population comprised speakers of Tamil and Telugu. Like other European surgeons of the day, Browne was involved in the Company's diplomacy with local powers. In 1693 he was sent to Arcot, a newly established regional headquarters of the Mughal Empire. There, Browne successfully treated the wound

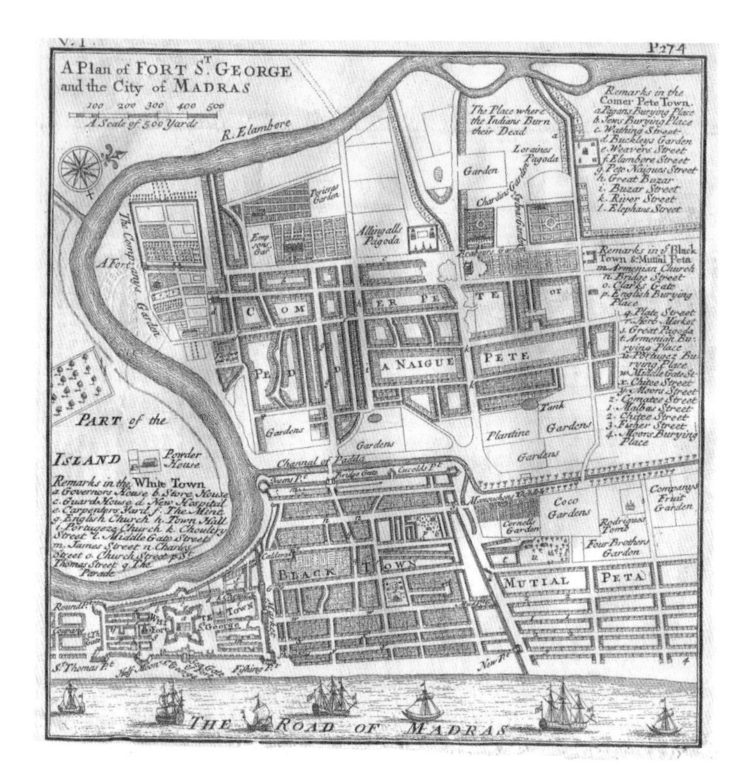

Left Herman Moll, 'A Plan of Fort St. George and the City of Madras', 1726. Edward Bulkley's garden, the plot formerly belonging to Browne, is marked with a 'd'.

of a general, Qasim Khan, who appointed him governor of six towns around Madras. The East India Company required him to refuse this position, but sent him on several other journeys to Arcot. Browne's collections demonstrate that he also spent time travelling with the Mughal military forces, where he would again have been employed as a surgeon and as an informal representative of the Madras settlement.

Browne's life in Madras was eventful. In 1693 he made a dramatic confession to the murder of James Wheeler, a member of the governing Council of Fort St George. Browne believed he had accidentally given Wheeler arsenic intended for another patient. However, after an investigation that included a post-mortem by Edward Bulkley, Browne's fellow surgeon at the Fort, Browne was acquitted. Browne found himself in trouble with the Governor and

Council of Fort St George on other occasions. In November 1695, he was arrested for drunkenly challenging another former Company surgeon, Richard Blackwall, to fight a duel and in 1697 he was fined for robbing and attacking a Mughal customs official. More seriously, in 1698 Company officials at Madras had to intervene in a dispute between Browne and one of his Mughal patients who had paid him a large sum to acquire medicines. Browne was accused of absconding with the money and the Company had to pay the debt on his behalf.

Throughout these adventures, Browne continued to collect local and exotic plants, cultivate them in his garden, and experiment with them for use in the hospital and in his private medical practice. He also bought other medical supplies in the local bazaars. He sent accounts of his experiments and observations along with his specimens to a range of correspondents in England. Most of the plants Browne collected were native to the areas around Madras. There are two interesting exceptions. The first is a specimen of *Iphiona scabra*, a woody plant of the daisy family with medicinal uses, native to the Arabian Peninsula and northeastern Africa. This was apparently gathered in Aden. In his account of the Sloane Herbarium, James Edgar Dandy assumed that Browne gathered the specimen himself on the way out to India. However, it is also possible that it was gathered by one of Brown's own collectors, who were often ship's surgeons. The second is a specimen of the Chinese rhubarb (*Rheum palmatum*). This specimen could in fact have been sent by Browne from Madras, since he wrote to Petiver in September 1698 that he was raising Chinese rhubarb in his section of the Company's garden.

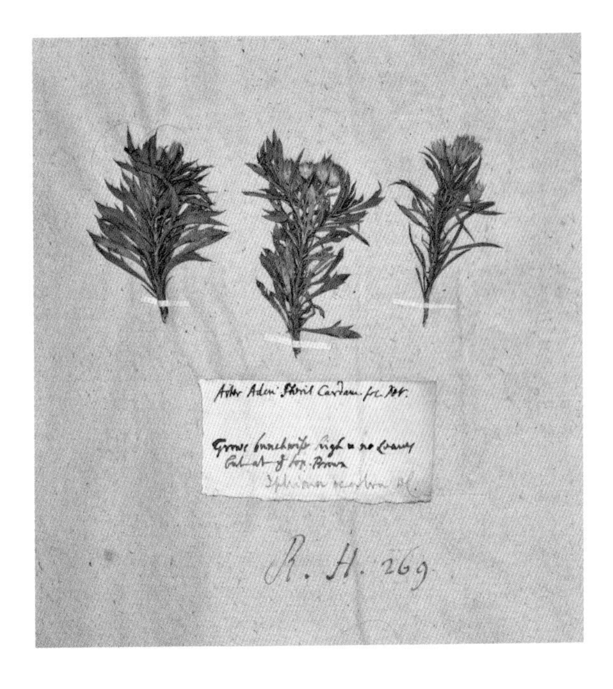

Below A specimen of *Iphiona scabra*, a medicinal plant from Aden, among Samuel Browne's collections in the Sloane Herbarium.

Over the course of their correspondence, Petiver sent Browne a number of books on botany, medicine and natural philosophy including the sixteenth-century writings of Prosper Alpinus and Garcia de Orta and the recent works by John Ray and Robert Hooke. He also sent the monumental *Hortus Malabaricus*, a contemporary account of Kerala's flora, compiled by the Dutch East India Company (VOC) surgeon Hendrik van Rheede tot Drakenstein. Browne was able to identify many of his specimens using these works, particularly the *Hortus Malabaricus*. He also passed on some books to other naturalists in his network. For example, Browne sent Ray's work on to Kamel in the Philippines, a gift that prompted Kamel to write to Ray in early 1699, initiating an ongoing correspondence between the two naturalists.

Through an extended collaboration with a Tamil-speaking physician and more occasional conversations with a speaker of Telugu, Browne was also able to supply many of the local names and uses of the plants he collected. Together with Petiver, Browne published a series of articles in the Royal Society's journal *Philosophical Transactions* between 1698 and 1703. These contain accounts of seven volumes of dried plants, 316 in total, that Browne had collected during 1696 and sent to London. The specimens are typically mounted and most are accompanied by a palm-leaf label which has the Tamil name written on it using a stylus that has been inked to make the text stand out. Browne's labels detailing the name, characteristics and uses of the plants are written on paper and pasted on to the pages containing the specimens. Finally, Petiver has added the relevant sections from the printed copies of *Philosophical Transactions of the Royal Society*. The volumes therefore provide a fascinating insight into the collaborative process behind the published accounts of the plants. Similar collaborative work is evident in other volumes in the Sloane collection.

Browne was dismissed as surgeon in January 1698, having refused an offer of a position as surgeon in Sutanuti, one of the three villages that would become Calcutta. Instead, he remained in Madras, taking on the position of assay master. Writing to Petiver that year, Browne noted that he planned to give up botany and had handed his section of the Company's garden over to his successor, Edward Bulkley. Nonetheless, Browne remained curious about natural philosophy, requesting a long list of books including works on architecture, medicine, physics and mathematics. In December 1698, Browne died and was buried alongside his son, also named Samuel Browne, in St Mary's churchyard at Fort St George. His wife Ann remarried and survived until 1715.

Browne's collections were remarkable for his time. They came close to the *Hortus Malabaricus* both in their extent and in terms of Browne's sustained collaboration with local people with medical knowledge. The demand for Browne's specimens and the interest they received from Petiver, Plukenet and others reveals the huge appetite for knowledge about exotic medicinal plants in seventeenth-century London.

Below Browne's specimen of *Buchanania cochinchinensis* commonly known as cherry-mango, with Browne's original hand-written notes on the left-hand folio, followed by the published version by Petiver and the vernacular name in Malabar 'Moruttan-chidde' inscribed on the palm leaf label on the right-hand folio.

GEORG JOSEPH KAMEL

SEBESTIAN KROUPA, DOMINGO MADULID AND
ESPERANZA MARIBEL AGOO

Opposite Kamel's specimen of 'Tchia', or the tea plant. In recognition of Kamel's contribution to the study of plants, the Swedish naturalist Carl Linnaeus coined the genus *Camellia* in 1753, which now includes the tea plant.

When it comes to dried plants, it is hard not to think of tea – especially in the British Isles. Enjoyed by hundreds of cultures around the world and second in consumption only to water, this beverage is commonly prepared from the dried leaves of the tea plant from the genus *Camellia*. Although the Sloane Herbarium holds numerous samples of *Camellia*, one of them tells a particularly intriguing story. In volume HS 165, we find a specimen that was collected by the Jesuit missionary Georg Joseph Kamel around the year 1700 in the Philippines. The similarity in names is not coincidental: the Swedish father of taxonomy Carl Linnaeus named the genus *Camellia* after Kamel in recognition of his contribution to the study of plants. Why was this Jesuit, stationed far away in Manila, bestowed with such an honour? How did his specimens end up in the Sloane Herbarium?

Kamel was born in 1661 in Brno in the present-day Czechia, which was then part of the Habsburg Monarchy. In its Catholic realms, education was monopolized by the Society of Jesus and Kamel thus attended the local Jesuit college. Young Kamel was especially curious about plants and upon finishing his secondary education, he joined the Society in 1682 and was apprenticed as an apothecary. Remedies at that time were largely plant-based and Kamel therefore extensively studied medicinal plants and their properties and learned to prepare different concoctions, poultices and ointments from roots, leaves and other plant parts.

In 1687, now as a qualified pharmacist, Kamel was selected for the Jesuit overseas missions to spread the gospel among non-Christians. His new home was to be the distant Philippine archipelago, then a colony under the Spanish flag. After more than a year-long voyage that took him across the Atlantic and the Pacific Oceans, Kamel arrived in Manila in 1688. As a trained pharmacist,

Tchia Sin: accedens frutex.

his task was to look after the health of his fellow Jesuits and of other local residents, colonists and indigenes alike. Due to his medical duties, Kamel devoted himself to the study of local and especially medicinal plants, specimens of which he began to collect and document to use in his practice.

Due to Kamel's diligence and growing expertise, his reputation soon spread beyond the Philippine Islands. By the late 1690s he began corresponding with two members of the Royal Society in London, the pharmacist James Petiver and the father of English natural history John Ray. Both Ray and Petiver worked under the patronage of Sir Hans Sloane. At the turn of the eighteenth century, Philippine natural riches were largely unknown to European scholars and Kamel's communications were the first comprehensive accounts of the local flora and fauna to reach Europe. His work was greatly valued by the learned community; for example, in his letter to Sloane, Ray wrote that 'I cannot but look upon it as an effect of Providence to stir up a man so well skilled in plants to apply himself to the inquisition, delineation, and description of the plants growing in those remote parts of the world, and giving an account of their virtues and uses'.

When sending plants between Manila and London, Kamel and his correspondents relied on merchant ships navigating the world's seas. Given the distance between the two cities, it usually took at least 10 months for a consignment to reach its destination. In these conditions, the most reliable means of communicating the information about plants was through dried specimens and images. Sending these shipments through commercial vessels was not without risks. For instance, when Kamel dispatched his account of the flora of the Philippines in 1698, 'the ship was assailed by pirates and the evidence of ten years of my labours, I fear, was lost in a single day'. Undaunted by this tragedy, Kamel resumed his work and managed to recreate his treatise on the Philippine flora just in time for it to be appended to Ray's masterpiece *Historia Plantarum* ('History of Plants', 1704). With the outbreak of the War of the Spanish Succession in 1701, which pitched Spain and England against each other, interruptions and delays continued to plague Kamel's communications with London. Kamel died of dysentery in 1706, but his English friends did not learn about his sorry fate until four years later.

Despite arduous circumstances, Kamel sent almost 2,000 descriptions, more than 800 images and nearly 250 specimens of plants to London. These items remained in Petiver's ownership until his death in 1718, when they were purchased by Sloane. Kamel's contribution to the Sloane Herbarium is the oldest surviving plant collection from the Philippines and consists of a wide range of plants: from seaweeds, ferns and herbaceous weeds to climbers, shrubs and trees.

Kamel's plants are of great historical and scientific value, as they provide a snapshot of the natural and cultural landscape of old Manila

Left Kamel's specimen of 'Bitanghol' (*Calophyllum inophyllum*), the source of tamanu oil, which is used in skin care to this day. Kamel recommended decoction prepared from fresh leaves for various skin problems, including warts, calluses and swellings.

during Spanish colonialism. Some of the trees, such as *Heritiera littoralis* and *Calophyllum inophyllum*, depict the vegetation on the banks of the Pasig River, a busy waterway that divided the local Filipino and Chinese communities from the Spanish walled city of Intramuros, where Kamel lived. Some of these trees served as a source of wood for small boats and ships. Many of Kamel's specimens include descriptions of their medicinal uses and indigenous names. These plants are representative of the diversity of herbs that Kamel used in his own medical practice and of his encounters with the different local communities. These cross-cultural interactions, alongside Kamel's own investigations and experiments, served as his main source of information about Philippine plants. One example is Kamel's specimen and illustration of greater galangal (*Alpinia galanga*). Kamel recorded the plant under its indigenous name 'Lancoas', recommended its roots especially for digestive problems and reported its indigenous use as a poultice to treat wounds caused by poisoned arrows. Today its tubers are known for their antibacterial and antiviral properties and are regionally used in cooking. Another example is a specimen of tree cotton (*Gossypium arboreum*), the dried leaves and bark of which Kamel endorsed for digestive issues and women's medical problems, respectively. Native to the Indian subcontinent, its presence in the Philippines points to the early modern networks of

Below Kamel's specimen of 'Tongon' (*Heritiera littoralis*). This tree lined the banks of the Pasig River, which divided the Spanish walled city of Manila from the Chinese and Filipino communities. Kamel wrote that its wood is so hard that it resists all iron instruments.

commerce and migration that brought the two regions together. Kamel's specimens therefore possess not only botanical, but also linguistic, anthropological, cultural and medicinal significance. Through the lens of Kamel's plants, we can recreate the bustling, cosmopolitan world of early modern Manila.

Kamel's specimens often included only leaves and lacked other, especially floral, parts. This makes the identification of his specimens extremely difficult. In these circumstances, Kamel's indigenous names often provide a useful tool for identifying the plants: for instance, greater galangal is today known as langkawas in the Philippines, which roughly corresponds to Kamel's 'lancoas'. It is unclear why Kamel focused predominantly on leaves. Perhaps he expected the specimens to work in tandem with his images and descriptions, which tend to detail other plant parts more frequently. However, the publication of images was an extremely expensive enterprise, especially due to the high costs of copperplates. Unfortunately, due to the lack of financial resources, Kamel's English friends never managed to publish his beautiful drawings. Without any point of visual reference, Kamel's work rapidly fell into oblivion. It was Linnaeus' decision to name the camellia in Kamel's honour that reminds us of the esteem that the Jesuit sparked among the greatest botanists of his era. Today, the tea plant stands as testimony to the curiosity and industry of the man who placed the Philippines on the map of science.

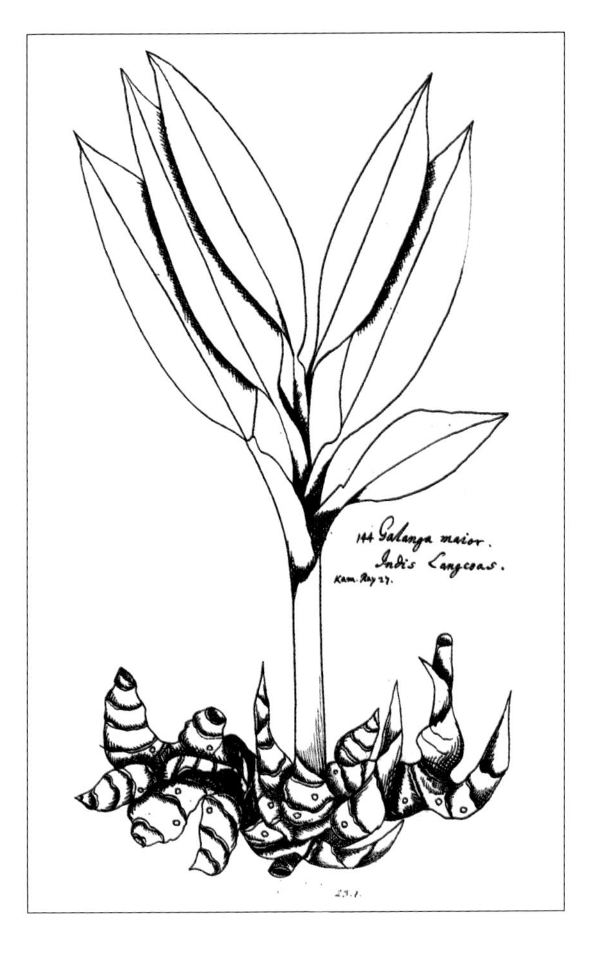

Above Kamel's drawing of 'Langcoas', or greater galangal (*Alpinia galanga*). Its roots, featured prominently in this image, were used for digestive problems and to prepare poultices to protect from wounds caused by poisoned arrows.

JAMES CUNINGHAME

RICHARD COULTON AND CHARLES E. JARVIS

James Cuninghame was the first European to make sustained natural history collections in China. Alongside Engelbert Kaempfer and Georg Kamel he ranks as the most significant contributor of specimens from East Asia to the Sloane Herbarium. Cuninghame visited China on at least two occasions as part of merchant ventures, spending six months in and around Amoy (Xiamen) in 1698, before living in the Chusan (Zhoushan) archipelago for two-and-a-half years between October 1700 and February 1703. The botanical material he accumulated there supplies a remarkably extensive and detailed record of the contemporary local ecologies. Moreover, Cuninghame's activity was by no means limited to China: botanical, entomological and conchological items survive in Sloane's collection from Indonesia, South Africa, the islands of the South Atlantic, and even the Kent coast. Before the age of plant hunters whose missions were more substantially planned and financed by sponsors back home, Cuninghame demonstrated a singularity of focus, attention to detail and (above all) the capacity for discovery that marks him out as a forerunner of the likes of Joseph Banks and Joseph Dalton Hooker.

Cuninghame was born in Scotland, probably in the mid-1660s, and appears to have matriculated as a medical student at the University of Leiden in 1686. By the mid-1690s he was engaged on commercial voyages to the East Indies as a ship's surgeon. En route he had begun collecting objects and information of natural historical interest as occasion permitted. During a period of furlough in London between two trips, Cuninghame and his specimens came to the attention of Sir Hans Sloane, probably following an introduction by James Petiver, an inveterate networker among itinerant men of medicine (although Cuninghame also had connections with other prominent scientists including John Woodward and Leonard Plukenet). Sloane and Petiver discovered in Cuninghame a sociable

Left A specimen of the dye plant 'Um-ki' or 'Wi-ki' (*Gardenia jasminoides*) collected in China by James Cuninghame in 1699 and accompanied by his original descriptive label.

and intelligent naturalist with energetic inclinations to sail again for distant parts. They became his patrons and correspondents, whom he entrusted with the safe-keeping of all he discovered during the course of his fieldwork. In recompense, Sloane proposed Cuninghame for election as a Fellow of the Royal Society, while Petiver was repeatedly to publish acknowledgements of the many natural rarities he had received thanks to 'the indefatigable Industry of my very worthy Friend Mr James Cuninghame Surgeon'. Indeed, both men championed Cuninghame's work in print, submitting papers to the *Philosophical Transactions of the Royal Society* that explicitly disseminated his research.

The plants that Cuninghame encountered during his visit to sub-tropical Amoy in 1698 ranged from tiny ferns to large trees, many of them never seen before in the West. However, his London patrons were especially eager to learn about species of commercial, medicinal or ornamental importance. Cuninghame's manuscript journal records in great detail the lengthy method of preparation of a pigment used by the Chinese to dye silk scarlet (a colour then difficult to achieve in Europe). Two botanical ingredients were essential, the seeds of 'Wi-ki' (gardenia) and the flowers of 'Ang-hoa' (safflower), their Chinese names characteristically transliterated among his notes. Cuninghame brought back to England dried specimens of both (*Gardenia jasminoides* and *Carthamus tinctoria*, respectively), as well as a locally drawn watercolour of each plant. He had also procured samples of two other components, 'black plum' and salt, which were incorporated within Sloane's Vegetable Substances collection. Cuninghame's gardenia specimens are

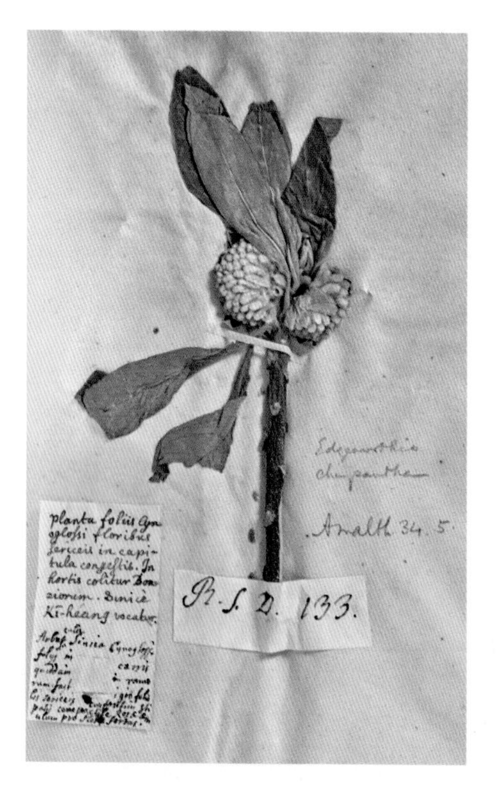

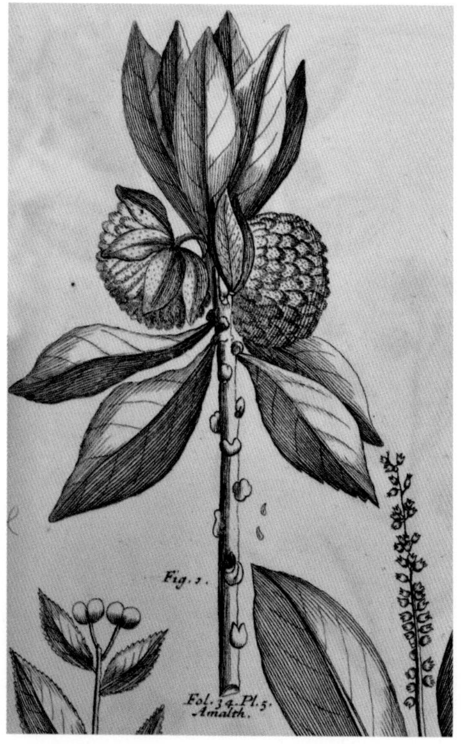

Below A specimen (left) of 'Ki-heang', (a species now known as *Edgeworthia chrysantha*), collected by Cuninghame in the Chusan archipelago with the engraving (right) prepared directly from it published by Leonard Plukenet.

present in three different volumes of Sloane's herbarium, reflecting their passage through the hands of Petiver, Plukenet and Sloane himself. Presumably the absence of raw materials in England explains why no attempt to replicate this Chinese dye-making process has been documented.

Two years later Cuninghame was again in China, travelling as surgeon aboard the *Eaton*, an East India Company vessel whose personnel were seeking to establish a trading post or 'factory' on the island of Chusan. While there Cuninghame managed to collect a considerable amount of carefully preserved material (both plant and animal). The plants were often accompanied by very detailed labels bearing long descriptions in Latin and records of vernacular names and the months of flowering and fruiting. From Petiver, Cuninghame had already received a list of Chinese plant desiderata gleaned from the few relevant publications in existence. One of the species was what Petiver called the 'candle tree' (*Triadica sebifera*), a generally useful and widely cultivated plant, the seeds of which yielded a wax deployed for candle-making. Specimens that Cuninghame sent to England survive in the herbarium volumes of both Petiver and Plukenet, who each published descriptions and figures of the flowers and fruits. This engaged a wider audience for the knowledge that Cuninghame was gleaning. Nearly 50 years later, based partly on Petiver's published account, Carl Linnaeus gave the binomial name *Croton sebiferum* to the species.

Above A living plant of *Edgeworthia* showing the inflorescence heads densely packed with flowers bearing long, silky hairs, as noted by Cuninghame on his specimen label ('floribus sericeis in capitula congestis').

Cuninghame had also been prompted by Petiver to research the tea-plant, in particular to solve the mystery of the distinction between black (or 'Bohea') and green tea, newly fashionable exotic drinks in London. In the Zhoushan archipelago Cuninghame had unprecedented access to observe the shrub (*Camellia sinensis*) growing – fine specimens sent to Petiver and Plukenet once again survive in the Sloane Herbarium – and to witness the commercial production of dried tea. His written account of this process was printed in the *Philosophical Transactions of the Royal Society* and explained that the various 'Sorts of Tea commonly carry'd to England are all from the same plant', making Cuninghame the first European to understand that the difference between black and green tea-leaves is a function of manufacture rather than biology. This insight was not wholly trusted by Western botanists until Robert Fortune's expedition to the Chinese interior almost 150 years later, but its accuracy indicates the value of Cuninghame's contributions to later scientific knowledge. A number of the published engravings that were based on his collections now serve as nomenclatural types for their corresponding Linnaean names. Nonetheless, other Cuninghame species described and illustrated by Plukenet were subsequently overlooked until much later, confirming a sense that Cuninghame was a botanist ahead of his time. This is the case for the ornamental flowering shrubs *Loropetalum chinense* (Chinese fringe flower), described by Robert Brown in 1818, and *Edgeworthia chrysantha* (paper bush), recognized by John Lindley only in 1846.

Cuninghame's legacy is above all one of scientific significance. However, the story of his overseas adventures is also a reminder of the excitements and the dangers of global travel during the period, not to mention the risks attendant on forging new trading relationships in remote intercultural contexts. En route to Amoy in 1698, his ship

Below Engraving of a flowering and fruiting stem of the 'candle tree' (*Triadica sebifera*), published by Leonard Plukenet and based on Cuninghame's specimens from Chusan.

Tuscan was impounded in the Canary Islands, and the enterprise was only permitted to proceed following the intercession of a resident Catholic priest with whom the surgeon corresponded in Latin. Far worse was to follow during Cuninghame's final long voyage between 1700 and 1709. In the first case, the East India Company project to found a permanent factory at Chusan – where he was to be resident surgeon – was a disaster, bringing the European merchants repeatedly into conflict (sometimes violently) with the Chinese authorities. The decision was taken to transfer the undertaking to the Vietnamese island of Pulo Condore (Côn So'n). This colony was first ravaged by disease, while ultimately its inhabitants were massacred by a combination of the mercenaries employed for their protection and forces loyal to the local Nguyen ruler. Cuninghame was spared but did not escape the region for a further two years, during which time he was kept under house arrest in Ba Ria. Upon his release he was reassigned to the English factory at Banjarmasin on Borneo, only for the settlement to be attacked and destroyed within a few weeks of his arrival. Eventually Cuninghame secured a passage home on the *Anna* but the ship never arrived in London. It is presumed to have been lost in the Indian Ocean during the spring of 1709.

Above One of Cuninghame's specimens of the 'candle tree' (*Triadica sebifera*), with a long descriptive label in his own hand recording the vernacular names for both the tree ('Kiu-tze') and the wax it produces ('Kiu-yeu').

ENGELBERT KAEMPFER

EDWIN ROSE

Engelbert Kaempfer (1651–1716), the son of a church pastor, was born in the town of Lemgo in northwestern Germany and went on to study medicine in Germany and Poland before graduating from Kraków. After several years in Königsberg and Uppsala, Kaempfer joined a Swedish diplomatic expedition in 1683 that travelled through Russia and Persia. By 1684 the Embassy had arrived in Isfhan, the Persian capital, where Kaempfer first started to collect the plants that can be found in the Sloane Herbarium.

In 1688 Kaempfer took employment with the Dutch East India Company (*Vereenigde Oostindische Compagnie* or VOC) as a surgeon and travelled through India, Ceylon (now Sri Lanka) and Java, at that time in the Dutch East Indies (Indonesia). The botanical component of Kaempfer's collection is now preserved in two volumes of the Sloane Herbarium. The first of these volumes contains 410 specimens on 111 folio sheets and was entitled by Matthew Maty, Principal Librarian of the British Museum, as '*Volumen plantarum in Japonia collectarum ab Engelberto Kaempfero M. D. annis 1691 & 1692. Addita sub finein plantae aliquot ab eodem in Persia & insula Ceylon repertae*', in short, collections of plants made by Engelbert Kaempfer from Japan in 1691 and 1692, in addition to those collected from Persia and the island of Ceylon.

Kaempfer found employment as a VOC surgeon in Japan in 1691, arriving in the Dutch enclave of Dejima, Nagasaki, on 26 September, where he remained until the end of October 1692. During this stay of slightly under 13 months, Kaempfer made vast collections of Japanese artworks, surgical equipment, porcelain, glassware, medicinal remedies, acupuncture pins, Buddhist icons, maps, drawings by Japanese artists and a large quantity of dried plants, many of which are now held in the Sloane Herbarium. These plants are backed on either Persian or Japanese paper, reflecting the

Left The specimen of *Camellia sinensis* Kaempfer collected and mounted on Japanese paper. Kaempfer noted phonetic transcriptions for the word 'Tea' next to the specimen.

original wrappings into which they were placed by Kaempfer, who tended to purchase locally made paper wherever he made port. The albums and notebooks in which Kaempfer preserved the specimens were disassembled by Sloane's curators and mounted into larger volumes.

One of the most notable plants collected by Kaempfer is a specimen of *Camellia sinensis*, the primary function of which is to produce tea. This is perhaps one of the oldest specimens of this species in existence. Kaempfer collected this specimen due to his interests in its potential medical qualities and the prominent position the tea camellia and tea itself held within Japanese society.

Left Kaempfer's copperplate illustration of the tea plant from his *History of Japan* (1727).

In his posthumous *The History of Japan* (1727), Kaempfer described the various processes used to cultivate tea and its social applications in a section entitled 'The Natural History of the Japanese Tea; with an accurate description of that Plant, its culture, growth, preparation and uses'. This description was illustrated with a fine copperplate image next to which Kaempfer gave the Chinese characters and the Latin translations of the Japanese plant names. Moving on from this, Kaempfer gave an account of 'what relates to the culture of the Tea shrub, beginning from the first planting of the seed'. After giving a detailed account of the processes of cultivating tea plants, Kaempfer described how it was harvested and processed, going through the different kinds of tea produced, from the 'Imperial Tea', made from smaller, younger leaves of the highest quality, through to those reserved for the lower classes:

> *The leaves of this gathering are sorted again, according to their size and goodness, into different classes, which Japanese call Itzhan, Niban and Sanban, that is the first, second and third, the last of which contains the corsest leaves of all, which are a full two months grown, and are the tea commonly drunk by the vulgar.*

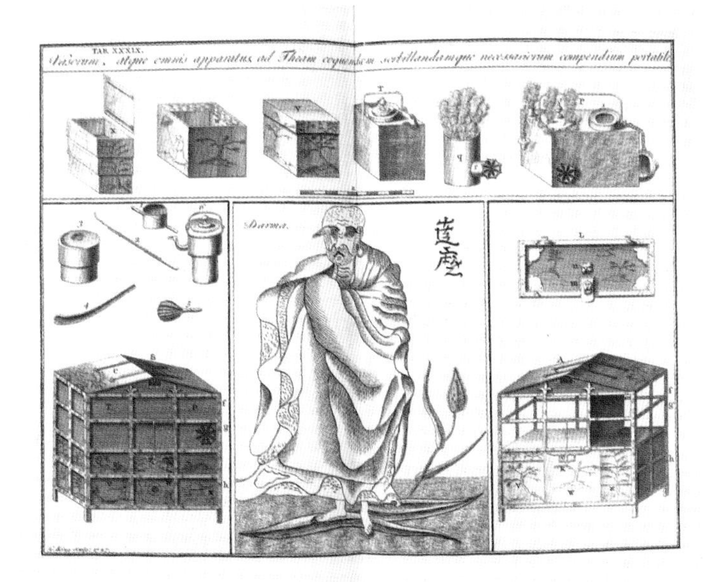

Left An illustration depicting the processes of making and grading different kinds of tea from Kaempfer's *History of Japan* (1727).

Kaempfer then discusses the preparation of the tea leaves in 'publick roasting-houses' where the tea is roasted, dried and then placed in airtight boxes. In producing these images and descriptions, Kaempfer brought a snapshot of Japanese life to contemporary European audiences, who had minimal knowledge about Japan.

Another botanical species Kaempfer collected is *Ginkgo biloba*, more commonly known as the ginkgo tree. Kaempfer is famed as the first person to have brought a specimen of this plant to Europe, which is now preserved in the Sloane Herbarium. This is the earliest specimen of ginkgo known to exist. The species was described by Kaempfer in his seminal publication *Amœnitatum Exoticarum Politico-Physico-Medicarum*, published in Lemgo in 1712. In this work Kaempfer systematically described the plants of Japan, many of which he collected illegally with the help of his interpreter Imamura Gen'emon during the annual pilgrimage to Edu (Tokyo) when VOC officials left the enclave of Dejima to pay homage before the Japanese court. In *Amœnitatum Exoticarum*, Kaempfer gave the Chinese characters, Latin transcriptions and a description of each plant he listed. Additionally, fine copperplate engravings were produced to represent certain species, based on 217 botanical illustrations in a folio volume entitled *Delineatio*

Plantarum Japonicarum, now held by the British Library. These had been produced in Japan under Kaempfer's instruction and found their way to Sloane along with the rest of the collection. The images only occasionally share similarities with the specimens from the Sloane Herbarium. Many of the specimens, as in the case of the ginkgo, are only fragmentary, suggesting that Sloane only received part of Kaempfer's botanical collection.

Sloane acquired Kaempfer's collection in 1723 from Kaempfer's nephew, Johann Hermann Kaempfer, an impoverished physician living in Hannover. To secure the collection, Sloane paid Kaempfer the hefty sum of £250 and raised a subscription to publish Kaempfer's manuscript entitled *History of Japan* (1727), which was translated by one of his curators, the Swiss naturalist Johann Casper Scheuchzer. The addition of Kaempfer's Japanese collection made Sloane's herbarium a truly global enterprise, capable of circulating knowledge on Japan across Europe. Information on Japan had previously been unavailable due to the isolationist policy of the Edo government.

The next individual who paid a significant amount of attention to Kaempfer's material was Daniel Solander (1733-82), a Swedish naturalist and former student of Carl Linnaeus. Solander had been employed by the British Museum in the 1760s to reclassify Sloane's collection according to the Linnaean system. The incomplete nature of the specimens made Solander's task particularly difficult. Many of the specimens, such as the ginkgo, are sterile, lacking flower and fruit, features that were essential for

Below The specimen of *Ginkgo biloba* Kaempfer collected in Japan between 1691 and 1692, with, below, the copperplate engraving of *Ginkgo biloba* from Kaempfer's *Amœnitatum Exoticarum* (1712).

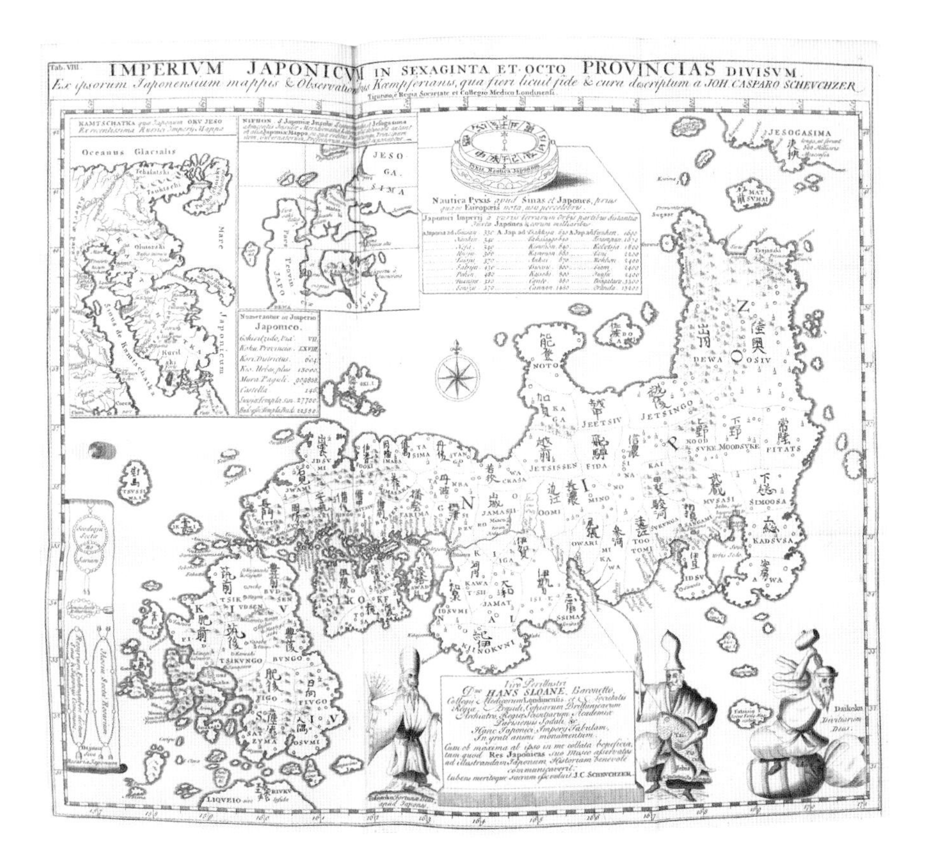

classifying species according to the Linnaean system. In one of his frequent reports to the Trustees of the British Museum, Solander complained that 'In general, the Specimens have been very good and compleat, except in Dr Kaempfer's Hortus Siccus, which seems to be made up of spare specimens'. In several cases, it seems that Solander received assistance in identifying Kaempfer's specimens from another Linnaean student, Carl Peter Thunberg, who visited London when returning from Japan in 1778. Kaempfer's specimens and descriptions formed the essential groundwork for Linnaeus and his students, such as Thunberg and Solander, to compile and publish Latin binomial names for the plants of Japan, a number of which are still in use today.

Above A map of Japan from Kaempfer's *History of Japan* (1727).

GEORGE HANDISYD

MARK CARINE AND TOD STUESSY

In 1692, in a letter to his friend the apothecary and physician Samuel Dale, Sir Hans Sloane described George Handisyd as '[M]y very good friend and ingenious gentleman…' Looking through Handisyd's collections today in the Sloane Herbarium, you are immediately struck by the unusually detailed labels, in Sloane's own hand, that accompany many of the specimens and by Sloane's notes detailing the species that Handisyd collected and where they came from. Sloane evidently held Handisyd in high regard and studied in great detail the specimens that he collected.

We know very little about Handisyd's life but two letters from Handisyd to Sloane, both written in 1692, provide some insight into his role as a ship's surgeon and his collecting activities. The first, dated 19 February 1692, was sent from the *Modena,* a cargo ship taking broadcloth to Mumbai and Surat in India. Having just embarked at the Port of London, Handisyd apologized to Sloane for not having 'time to take my leave of you and others of my good friends' and explained that he had been '…sent for in all heast on bord to on on[e] of our maites who is ill of a putrid fever'. Later that year, on 27 July, he wrote from Cape Bona Speranze (the Cape of Good Hope in South Africa). The letter is torn but his collecting ambitions are certainly made clear: 'I hope to increase your stocke especialy to pla [torn] having maid it my business ever since we arivd at this port to looke out for them and also imployd others for y^{t} intent'.

Despite his best intentions, no specimens from that voyage appear to be in the Sloane Herbarium today. Rather, it is the specimens that Handisyd collected earlier, in around 1690, that drew Sloane's attention. Those specimens include plants collected in New England, Hispaniola and Barbados, but also among them are plants collected in the Straits of Magellan, and in the eastern Pacific

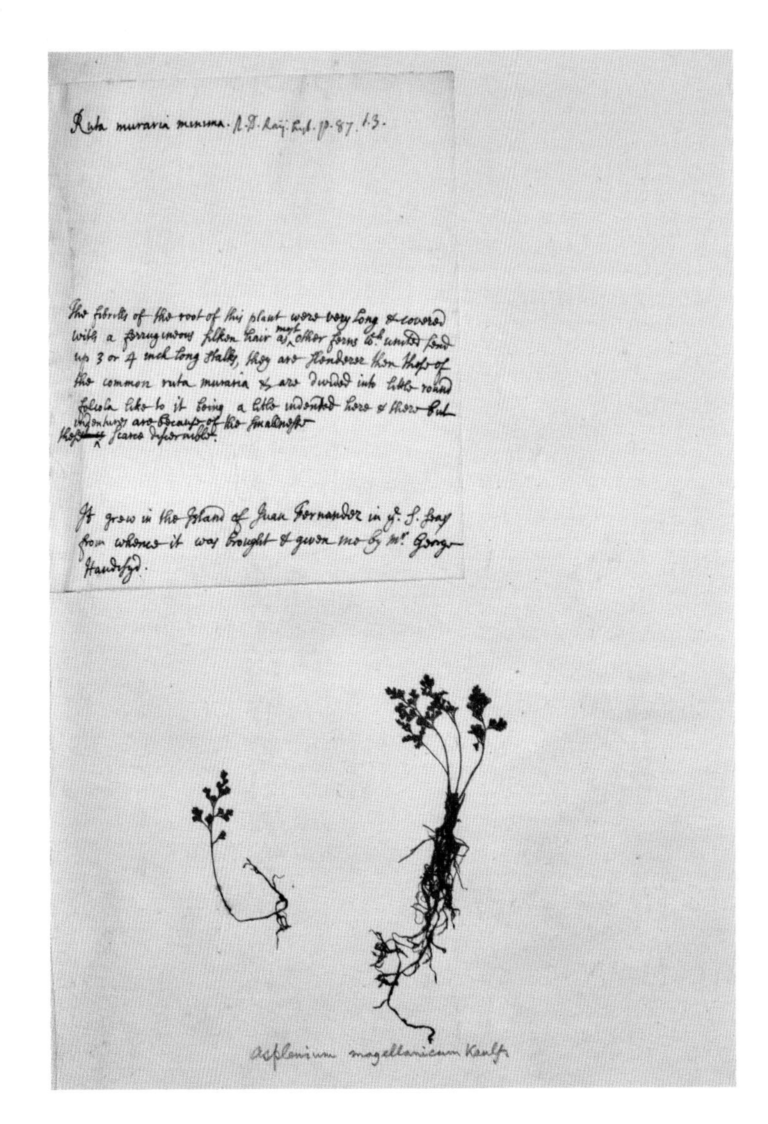

Left Handisyd's specimen of the fern *Asplenium dareoides*, collected on the Juan Fernández Islands and accompanied by detailed notes in Sloane's hand.

and it is those collections that stand out as Handisyd's unique contribution to Sloane's herbarium.

Specimens collected by Handisyd are in the collections assembled by both Sloane and William Courten. His specimens are the only specimens from temperate South America in Sloane's herbarium and they appear to be the earliest surviving collections from the region anywhere in existence. They were made far from established areas of British trade and influence. Chile was, at that time, a Spanish Vice-Royalty and the Treaty of Madrid, signed in 1670, meant that England guaranteed not to challenge Spain's commercial monopoly in America in return for Jamaica and the Cayman Islands. Whilst that was, at least, the official position, British ships continued to visit the region and by the 1680s British acts of piracy in South America had become common place.

Many of Handisyd's specimens have locality details that allow us to build a picture of where he collected. We find for example specimens collected '[F]rom the first narrow in ye Magellan str.' and the Straits of Magellan was the focus for much of his collecting. He collected on Elizabeth's Island [Isla Isabel], close to the northern shore and at Port Famine, Batchelor's River and Cape Quad, all located on the northern side of the straits. On the southern shore, eight plants were collected from 'Port Gee' in western Tierra del Fuego, a locality that today cannot be traced.

Handisyd collected lichens, mosses and ferns as well as flowering plants and many of the species he collected occur only in temperate South America. His specimens would have provided an insight into a flora entirely new to Sloane whose first-hand experience of American plants was gained in the tropical Caribbean region.

Among Handisyd's collections were specimens of *Drimys winteri*, commonly known as winter's bark or canelo. Those specimens enabled Sloane to establish the true identity of this useful but enigmatic plant. Over one hundred years earlier, in 1578, Francis Drake had passed through the Straits of Magellan during his circumnavigation of the globe. It was, to say the least, a difficult passage. Violent storms destroyed one of his three

Below Handisyd's specimens of *Drimys winteri*, the source of 'Cortex winterianus', allowed Sloane to determine the identity of this species, the bark of which had been first taken to Europe by Captain Wynter in 1578.

ships (the *Marigold)* and caused another (the *Elizabeth)* to turn back and return to England, leaving only the *Pelican* (later renamed the *Golden Hind*) to complete the voyage. In the straits, they discovered that an infusion made of the bark of *D. winteri* was an effective remedy for scurvy. The *Elizabeth* returned to England with bark from this tree and it was named *Cortex winterianus* by the eminent French botanist Carolus Clusius after John Wynter, the ship's captain. However, with no herbarium specimens with which to ascertain its diagnostic features, the identity of this species remained unresolved and apothecaries often sold the bark of another species under the same name.

Above Sloane's paper on *Cortex winterianus* published in the *Philosophical Transactions of the Royal Society* in 1693 included an illustration drawn from one of the specimens that Handisyd collected (see opposite, bottom left specimen).

Examination of Handisyd's specimens of this plant led Sloane to write a paper entitled 'An Account of the true Cortex Winterianus and the Tree that bears it'. It provided a description of *Drimys winteri* and detailed how Handisyd had used the plant, not only for curing scurvy but also for treating members of the crew who fell ill '...by eating a poisonous sort of Seal in those Parts called a Sea-Lion...' Sea lions have been used as a source of food and so the cause of this poisoning is not clear. Sloane's paper also included an illustration that is clearly based on one of Handisyd's specimens.

The labels that accompany Handisyd's plants indicate that he also collected a handful of specimens in the Pacific – from the 'Island of Mucho' (Isla Mocha), close to the Chilean coast and from the Juan Fernández Islands, located 670 km (about 400 miles) west from mainland Chile. The latter is a small, remote, and sparsely populated archipelago that came to popular attention in the early 1700s after Alexander Selkirk, the inspiration for Robinson Crusoe, was marooned there for four years. Selkirk was a privateer, a profession typical of many who visited the archipelago during that period.

Handisyd's specimens from the Juan Fernández Islands are in many respects unremarkable: just two fern species were collected, both of which are widespread in the archipelago and neither of which is unique to the Juan Fernández Islands. His specimens are, nevertheless, the first collections from this isolated archipelago. Indeed, no other collections would be made from the Juan Fernández Islands for more than 130 years, until the visit of Mary Graham in 1822.

A critical review of Handisyd's collections undertaken in the 1970s by the Chilean botanist Hugo Gunckel highlighted a number of discrepancies between the localities recorded on the specimens by Sloane and the known distributions of species collected. For example, Sloane's label suggests that the specimen of *Heliotropium stenophyllum* was collected in the Straits of Magellan whereas in reality it is a species that is restricted to semi-desert coastal areas much farther north. One specimen is very honestly labelled by Sloane 'I do not remember any thing Mr Handisyd said of this it seems to be a fern'. The labelling of specimens, by Sloane, and far removed in space and time from their point of collection meant that errors would have been almost inevitable.

Gunckel noted that *Heliotropium stenophyllum* is common around Coquimbo, a port where the *Wellfare,* an English privateer captained by John Strong, is known to have called in 1690. Indeed, Handisyd's itinerary, encompassing the Straits of Magellan, Isla Mocha, and the Juan Fernnádez Islands, closely matches that of the *Wellfare* leading Gunckel to suggest that Handisyd was most likely on board that ship.

Strong hoped to lead the *Wellfare* to establish trade in the Pacific and to salvage sunken Spanish silver. In both objectives he failed but his expedition was responsible for the first recorded landing on the Falkland Islands. Blown off course heading south along the Atlantic coast of Patagonia, the *Wellfare* landed on the Falklands on 27 January 1690 and the islands were named after the English Navy treasurer.

A locality provided on several of Handisyd's specimens is 'Port Falkland' listed as being in the Straits of Magellan. It is a locality that cannot be traced in the Straits. Gunckel suggested that it may actually refer to the Falkland Islands. A number of the species recorded from 'Port Falkland' can be found in the Falklands today, including *Gunnera magellanica* but this is not the case for all. Furthermore, in the case of the almond flower (*Luzuriaga marginata*) the label indicates that it grew 'in the woods by the mountains on Port Falkland'. There are no native trees on the Falklands. If Handisyd did collect on the islands at least some specimens have been mislabelled.

Today Handisyd is remembered for making the first botanical collections from Chile. That he may also have been the first to collect on the Falkland Islands, long before the next recorded collections made in 1789, is an intriguing possibility. As a ship's surgeon in the South Seas, almost certainly aboard an English privateer, plants and their uses were integral to Handisyd's work. As a collector, the way in which provenance information was recorded fell well short of modern standards but the specimens he made, far from established British interests, formed a unique and highly valued component of Sloane's herbarium.

Left The red fruits of *Gunnera magellanica* growing in the Falkland Islands today. Handisyd collected this species at 'Port Falkland', a locality that cannot been traced. If Handisyd were on the *Wellfare* as has been suggested, his specimen may have been collected in the Falkland Islands.

Timeline for the people and main events

1627: John Ray born, Black Notley, Essex.

1630: Mary Somerset, Duchess of Beaufort, born.

1642: William Courten, Robert Uvedale and Leonard Plukenet are born; all three are subsequently educated at Westminster School.

1651: Engelbert Kaempfer born, Lemgo, Germany.

1656: Bruno Tozzi born, Montevarchi, Italy.

1658: James Petiver born, Hillmorton, Rugby.

1659: William Sherard born, Bushby, Leicestershire.

1660: Sir Hans Sloane born, Killyleagh, Co. Down; Edward Lhwyd born, Loppington, Shrophsire; John Ray publishes his first botanical work – an account of the flora of Cambridgeshire.

1661: Georg Joseph Kamel born in Brno, Czechia.

1662: John Ray embarks on a tour of continental Europe, returning to England in 1665; Richard Richardson born in North Bierley, Yorkshire; Adam Buddle born in Deeping St James, Leicestershire.

?1674: John Lawson born in London.

1669: David Krieg born in Annaberg, Germany.

1679: Pier Antonio Micheli born in Florence, Italy; Sloane leaves Ireland for London where he trains as an apothecary.

1681: George London establishes the Brompton Park Nursery on the site of today's Victoria & Albert Museum.

1682: Mark Catesby born at Castle Hedingham, Essex; Lhwyd makes his first botanical tour of North Wales; George London is commissioned to design a garden for Lord Weymouth at Longleat.

1683: Sloane travels to France, where he meets Courten, studies botany in Paris and Montpellier and takes a medical degree at the University of Orange; he returns to England in 1684. Kaempfer travels to Russia and Persia. Jaun Salvador i Riera born in Barcelona.

1684: The husband of Mary Somerset, Duchess of Beaufort, inherits Badminton House, Gloucestershire, where the Duchess creates a 'glorious garden'.

1686: Ray publishes the first volume of his *Historia Plantarum,* subsequently used by Sloane to catalogue his herbarium; the remaining two volumes are published in 1688 and 1704.

1687: Sloane travels to Jamaica as physician to the Governor, the Duke of Albemarle, returning to England in 1689.

1688: Kamel arrives in Manilla, Philippines; Kaempfer travels to India, Sri Lanka and Java.

?1690: George Handisyd travels through the Straits of Magellan and into the Pacific.

1691: Philip Miller born in Deptford or Greenwich. Plukenet publishes the first of three volumes of his *Phytographia.* Kaempfer appointed VOC Surgeon in Japan, a post he holds until 1692.

1693: Edward Bartar returns to the Gold Coast having met Petiver and Plukenet in London; he subsequently sends plant specimens to Petiver.

1695: William Houstoun is born in Houston, Renfrewshire. Petiver publishes the first of six parts of his *Museum Petiverianum.*

1697: Samuel Browne sends seven books of plants to the East India Company in London from Fort St George (modern day Chennai) for examination by Petiver; he dies a year later. Krieg travels from Riga to London and a year later to Maryland to collect plants.

1698: Kamel's account of the flora of the Philippines is lost to pirates. James Cuninghame makes his first voyage to China; he returns for a further 2 ½ years between 1700–1703.

1699: John Bartram born in Pennsylvania, USA.

1700: Lawson travels to Carolina where he is employed as a surveyor, rising to become Surveyor General in 1708.

1702: Courten's herbarium is acquired by Sloane after Courten's death in this year.

1703: Sherard appointed the Levant Company Consul in Smyrna; a post he held until he returned to England in 1717.

1706: Kamel dies in Manilla, Philippines. Plukenet dies and his herbarium mostly passes into the hands of Dr Moore, Bishop of Norwich.

1707: Sloane publishes the first volume of *Natural History of Jamaica.* Buddle completes an unpublished English flora based on his herbarium.

1709: Lhwyd dies in Oxford. Cuninghame is thought to have died on a ship returning to England from Borneo.

1710: Lawson returns to North America after a short visit to London; a year later, he is tried and killed by the Tuscarora people.

1712: Sloane purchases the Manor of Chelsea. Catesby travels to Virginia where he remains until 1719. Salvador i Riera botanises in Majorca and Minorca and sends specimens to Petiver.

1713: Tozzi starts to correspond with Petiver, also sending him specimens.

1714: The Duchess of Beaufort dies in Chelsea at the age of 85 and her herbarium is acquired by Sloane.

1715: Adam Buddle dies in London; Sloane acquires his herbarium.

1716: Salvador i Riera embarks on an eight-month collecting trip around Spain and Portugal.

1718: Petiver dies in London; Sloane, a pall bearer at his funeral, subsequently purchases his herbarium and other collections 'for a considerable sume'.

1722: The Chelsea Physic Garden is conveyed by Sloane to the Society of Apothecaries and Philip Miller is appointed its Gardener, a post he holds until 1770; Sloane is one of the sponsors of Mark Catesby's four-year exploration of the Carolinas, Georgia, Florida and the Bahamas; Robert Uvedale dies in Enfield.

1723: Sloane purchases Kaempfer's herbarium for £250.

1726: Salvador i Riera dies in Barcelona, Spain.

1727: Sloane elected president of the Royal Society, a post he would hold until 1741.

1728: Houstoun travels to the West Indies, returning a year later; Bartram purchases the land on which he establishes his garden; Sherard dies.

1729: Micheli publishes his *Nova Plantarum Genera;* Catesby produces the first 20 hand-coloured plates of his *Natual History of Carolina, Florida and the Bahama Islands*.

1730: Houstoun travels to Jamaica and Mexico, employed by the South Seas Company; he dies in Jamaica in 1733.

1731: Philip Miller publishes the first edition of his *Gardener's Dictionary*, dedicated to Sloane.

1733: Specimens are sent to Sloane by General James Edward Ogelthorpe from the Colony (now State) of Georgia that he founded this year; Daniel Solander born in Piteå, Sweden.

1737: Micheli dies in Florence, Italy.

1740: Sloane acquired Uvedale's herbarium after the death of his widow this year.

1742: Bartram sends plants to Sloane; a second consignment is sent in 1743.

1743: Tozzi dies in Vallombroa, Italy.

1749: Catesby dies in London.

1753: Sloane dies on 11th January at the Manor House, Chelsea; his herbarium, together with the rest of his collection, is purchased by the nation and becomes the foundation collection of the British Museum.

1763: Solander is employed as an Assistant at the British Museum to catalogue the natural history collections, including the Sloane Herbarium.

1771: Philip Miller dies; his personal herbarium is subsequently purchased by Joseph Banks in 1774; Richard Richardson dies.

1777: John Bartram dies in Philadephia, USA.

1782: Solander dies at the home of Joseph Banks in Soho Square, London.

Bibliography

Introduction

Dandy, J.E. (1958), *The Sloane Herbarium: An Annotated List of the Horti sicci Composing It; With Biographical Details of the Principal Contributors*. British Museum (Natural History), London.

Delbourgo, J. (2017), *Collecting the World: the Life and Curiosity of Hans Sloane*. Allen Lane, London.

Edward Bartar

Bosman, W. (1705), *A New and Accurate DESCRIPTION of the Coast of Guinea*. London.

Law, R. (ed.) (2007), *The English in West Africa, 1691–1699: the local correspondence of the Royal African Company of England, 1681–1699, Part 3.* Oxford University Press, Oxford.

Murphy, K.S. (2013), Collecting slave traders: James Petiver, natural history, and the British slave trade. *William and Mary Quarterly*, 70 (4): 637–670.

Petiver, J. (1695–1703), *Musei Petiveriani Centuria*. Samuel Smith & Benjamin Walford, London.

Petiver, J. (1706), *Gazophylacii Naturæ & Artis* [decas sexta]. Christopher Bateman, London.

Yeboa Daaku, K. (1970), *Trade and Politics on the Gold Coast, 1600–1720*. Clarendon Press, Oxford.

John Bartram

Anon. (1928), The Disownment of John Bartram. *Bulletin of Friends Historical Association*, 17(1): 16–22.

Darlington, W. (1849), *Memorials of John Bartram and Humphrey Marshall with Notices of their Botanical Contemporaries.* Lindsay & Blakiston, Philadelphia.

Douglas, S. (2015), Dr John Fothergill: significant donor. In: M. Campbell, E.G. Hancock and N. Pearce (eds.), *William Hunter's World: the Art and Science of Eighteenth-century Collection*. Ashgate Publishing, Farnham, pp.165–176.

Kastner, J. (1978), *A World of Naturalists*. John Murray, London.

Laird, M. and Bridgman, K. (2014), American roots: techniques of plant transportation and cultivation in the early Atlantic world. In: P.H. Smith, A. Meyers and H.J. Cook (eds.), *Ways of Making and Knowing: the Material Culture of Empirical Knowledge*. University of Michigan Press, Ann Arbor, pp.170–172.

Driver, F. and Martins, L. (eds.), (2005), *Tropical Visions in an Age of Empire*. University of Chicago Press, Chicago.

Swem, E.G. (ed.), (1957), *Brothers of the Spade: Correspondence of Peter Collinson, of London, and of John Custis, of Williamsburg.* Barre Gazette, Barre, MA.

West, F. and Bartram, J. (1955), John Bartram and Slavery. *The South Carolina Historical Magazine*, 56(2): 115–119.

Wulf, A. (2009), *The Brother Gardeners: Botany, Empire, and the Birth of an Obsession*. Windmill, London.

Reverend Adam Buddle

Buddle, B.M. (2008), Moss-cropper extraordinaire: the Rev. Adam Buddle (1662–1715), *The Linnaean*, 24 (4): 13–19.

Dandy, J.E. (1958), *The Sloane Herbarium: An Annotated List of the Horti sicci Composing It; With Biographical Details of the Principal Contributors*. British Museum (Natural History), London, pp.102–108.

Dillenius, J. (1741), *Historia Muscorum*. Theatro Sheldoniano, Oxford.

Houston, W. (1781), *Reliquiae Houstounianae*. London.

Linnaeus, C. (1753), *Species Plantarum*, vol 1. Laurentius Salvius, Stockholm, p.112.

Noltie, H.J. (2008), On the name *Buddleja*: a supplement to B.M Buddle's 'Moss-cropper extraordinaire', *The Linnaean*, 24 (4):19–24.

Ray, J. (1686–1704), *Historia Plantarum*. London.

Ray, J. (1724), *Synopsis Methodica Stirpium Britannicarum*, edn. 4 , J. Dillenius (ed.). Gulielmi & Joannis Innys, London.

Sloane, H. (1725), *A Voyage to the Islands Madera, Barbados, Nieves, S. Christophers and Jamaica*, vol 2. London.

Mark Catesby

Catesby, M. (1754), *The Natural History of Carolina, Florida and the Bahama Islands*. Printed for C. Marsh and T. Wilcox. London.

Frick, G.F. and Stearns, R.P. (1961), *Mark Catesby: the Colonial Audubon*. University of Illinois Press, Urbana, Illinois.

Harris, S.A. (2015), *The plant collections of Mark Catesby in Oxford*. In: E.C. Nelson and D.J. Elliott (eds.), *The Curious Mister Catesby*. The University of Georgia Press, Athens, Georgia, pp.173–188.

Laird, M. (2015), Mark Catesby's plant introductions and English gardens of the eighteenth century. In: E.C. Nelson and D.J. Elliott (eds.), *The Curious Mister Catesby*. The University of Georgia Press, Athens, Georgia, pp.265–280.

McBurney, H. (1997), *Mark Catesby's Natural History of America: the Watercolours from the Royal Library, Windsor Castle*. Merrell Publishers Ltd., London.

McMillan, P.D. and Blackwell, A.H. (2013), The vascular plants collected by Mark Catesby in South Carolina: combining the Sloane and Oxford herbaria. *Phytoneuron*, 2013–73:1–32.

McMillan, P.D. et al. (2013), The vascular plants in the Mark Catesby collection at the Sloane Herbarium, with notes on their taxonomic and ecological significance. *Phytoneuron*, 2013–7: 1–37.

Nelson, E.C. and Elliott, D.J. (2015), *The Curious Mister Catesby*. The University of Georgia Press, Athens, Georgia.

William Courten

Gibson-Wood, C. (1997), Classification and value in a seventeenth-century museum. *Journal of the History of Collections*, 9: 61–77.

Kusukawa, S. (2017), William Courten's lists of 'things bought' from the late seventeenth century. *Journal of the History of Collections*, 29: 1–17.

Jarvis, C.E. and Cooper, J.H. (2014), Maidstone's woodpecker – an unexpected bird specimen in the herbarium of Sir Hans Sloane. *Archives of Natural History*, 41: 230–39.

James Cuninghame

Coulton, R. (2010), *Natural History and Medical Writing, volume II of Tea and the Tea-Table in Eighteenth-Century England*, 4 vols, M. Ellis (ed.). Pickering and Chatto, London.

Ellis, M. et al. (2015), *Empire of Tea – the Asian Leaf that Conquered the World*. Reaktion Books, London.

Jarvis, C.E. et al. (2014), *Gardenia jasminoides* – a traditional Chinese dye plant becomes a garden ornamental in Europe. *Curtis's Botanical Magazine*, 31: 80–98.

Jarvis, C.E. et al. (2015), The collecting activities of James Cuninghame FRS on the voyage of *Tuscan* to China (Amoy) between 1697 and 1699. *Notes & Records Royal Society*, 69: 135–153.

Kilpatrick, J. (2007), *Gifts from the Gardens of China*, Chapter 3. The First Collector. Frances Lincoln, London.

Santos-Guerra, A. et al. (2011), Late 17th century herbarium collections from the Canary Islands: the plants collected by James Cuninghame in La Palma. *Taxon*, 60: 1734–1753.

George Handisyd

Gunckel, H. (1971), Las primeras plantas herborizadas en Chile. *Anales del Instituto de la Patagonia*, 21: 134–141.

Lane, K. (2016), *Pillaging the Empire: Global Piracy on the High Seas, 1500–1750*, 2nd edn. Routledge, New York.

William Houstoun

Johnston, E.D., (1941), Dr. William Houstoun, botanist. *The Georgia Historical Quarterly*, 25(4): 325-339.

Rose, E.D., (2018), 'Natural history collections and the book: Hans Sloane's *A Voyage to Jamaica* (1707–1725) and his Jamaican plants', *Journal of the History of Collections*, 30 (1): 15–33

Georg Joseph Kamel

Letter from Georg Joseph Kamel to John Ray, 3 January 1699, Manila (British Library, Sloane MS 4062, f. 292v).

Letter from John Ray to Hans Sloane, 14 August 1700, Black Notley (British Library, Sloane MS 4038: f. 49).

Kamel, G. J., (1704), Historia stirpium insula Luzonis et reliquarum Philippinarum. In: John Ray, *Historiae PlantarumTtomus Tertius.* Samuel Smith and Benjamin Walford, London, pp. 1–96.

Kroupa, Sebestian (2016), *Ex Epistulis Philippinensibus*: Georg Joseph Kamel SJ (1661–1706) and His Correspondence Network. *Centaurus*, 57: 229–259.

Kroupa, S. (2019), *Georg Joseph Kamel (1661–1706): A Jesuit Pharmacist at the Frontiers of Colonial Empires*, PhD Thesis, University of Cambridge.

Madulid, D. (2001), *A Dictionary of Philippine Plant Names*, 2 vols. Bookmark Inc., Makati City.

David Krieg

Frick, G.F., Reveal, J.M., Broome, C.R. and Brown, M.L. (1987), Botanical explorations and discoveries in colonial Maryland, 1688-1753. *Huntia*, 7: 5-59.

Tering, A. (2015), Contacts in natural sciences between Riga and England in 1660–1710. *Ajalooline Ajakiri*, 1/2 (151/152): 39–83.

John Lawson

Lawson, J. and Lefler, H.T. (1967), *A New Voyage to Carolina. Edited with an introduction and notes by Hugh Talmage Lefler.* University of North Carolina Press, Chapel Hill.

Bellis, V. (2009), John Lawson's plant collections from North Carolina 1710-1711. *Castanea*, 74 (4): 376–389.

Blackwell, A.H., McMillan, P.D. and Blackwell, C.W. (2014), John Lawson's plant collections, Virginia and North Carolina 1710–1711. *Phytoneuron,* 2014–94: 1–23.

Mathewes, P. (2011), John Lawson the Naturalist. *The North Carolina Historical Review*, 88 (3): 333–348.

Edward Lhwyd (Lhuyd)

Camden, W. (1695), *Camden's Britannia, Newly Translated into English: with large additions and improvements*. A. Swalle for Edmund Gibson, London.

Gunther, R.T. (1945), *Early Science in Oxford, Vol. XIV, Life and Letters of Edward Lhwyd.* Oxford University Press, Oxford.

Ray, J. (1690), *Synopsis Methodica Stirpium Britannicarum.* Sam Smith, London.

George London

Bauhin, G. (1620), *Prodromus theatri botanici.* Joannis Regis, Basle.

Dandy, J.E. (1958), *The Sloane Herbarium: An Annotated List of the Horti sicci Composing It; With Biographical Details of the Principal Contributors.* British Museum (Natural History), London, pp. 157–159.

London, G. and Wise, H. (1713), *The Retir'd Gard'ner ...* Vol. I. Being a translation of Le Jardinier Solitaire ... Containing the methods of making ... a fruit and kitchen-garden ... Vol. II. containing the manner of planting ... all sorts of flowers, ... being a translation from the Sieur L. Liger. The whole revis'd, with alterations and additions, etc. London.

Morison, R. (1672), *Plantarum Umbelliferarum Distributio Nova.* Theathro Sheldoniano, Oxford.

Morison, R. (1715), *Plantarum Historiae Universalis Oxoniensis.* Theathro Sheldoniano, Oxford.

Pier Antonio Micheli

Jarvis, C. E., (2016), Pier Antonio Micheli (1679–1737) and Carl Linnaeus (1707–1778). *Webbia, 71*(1): 1–24. https://doi.org/10.1080/00837792.2016.1147210

Nepi, C. (2009,. L'Erbario Micheli-Targioni. In: Raffaelli, M. (ed.) *Il Museo di Storia Naturale dell'Universitá degli Studi di Firenze, Vol. II. Le Collezioni Botaniche*, Firenze University Press, Firenze, pp. 85–99.

Philip Miller

Britten, J. (1913), Philip Miller's plants. *Journal of Botany, British and Foreign,* 51: 132–135.

Le Rougetel, H. (1990), *The Chelsea Gardener: Philip Miller, 1691–1771.* British Museum (Natural History), London.

Stearn, W.T. (1971), Philip Miller and the plants from the Chelsea Physic Garden presented to the Royal Society of London, 1723–1796. *Transactions of the Botanical Society of Edinburgh,* 41(3): 293-307.

Stearn, W.T. (1974), Miller's Gardeners dictionary and its abridgement. *Journal of the Society for the Bibliography of Natural History*, 7(1): 125–141.

Stungo, R. (1993), The Royal Society Specimens from the Chelsea Physic Garden 1722–1799. *Notes and Records of the Royal Society of London*, 47(2): 213–224.

James Petiver

Coulton, R. (2012), The Darling of the Temple-Coffee-House Club: Science, Sociability and Satire in Early Eighteenth-Century London. *Journal for Eighteenth-Century London*, 35: 43–65.

Coulton, R. & Jarvis, C.E. (eds), (2020), Remembering James Petiver (c. 1663–1718). *Notes & Records Royal Society,* 74 (Special Issue), in press.

Hunt, A. (2018), Under Sloane's shadow: the archive of James Petiver. In: *Archival Afterlives: Life, Death and Knowledge-Making in Early Modern British Scientific and Medical Archives*, V. Keller, A. M. Roos and E. Yale (eds). Brill, Leiden, pp. 194–222.

Jarvis, C.E. (2018), Take with you a small Spudd or Trowell: James Petiver's Directions for Collecting Natural Curiosities, in: *Naturalists in the Field. Collecting, Recording and Preserving the Natural World from the Fifteenth to the Twenty–First Century,* A. MacGregor (ed.). Brill, Leiden and Boston, pp. 212–239.

Murphy, K. (2013), Collecting slave traders: James Petiver, natural history, and the British slave trade. *William and Mary Quarterly*, 70: 637–670.

Stearns, R.P. (1952), James Petiver: promoter of natural science. *Proceedings of the American Antiquity Society, 62*: 243–365.

Leonard Plukenet

Boulger, G.S. (1900), Some manuscript notes by Plukenet. *Journal of Botany, British and Foreign*, 38: 336–338.

Jackson, B.D. (1894), Leonard Plukenet. *Journal of Botany, British and Foreign,* 32: 247–248

Trimen, H. and Thiselton-Dyer, W. H. (1869), *Flora of Middlesex : a topographical and historical account of the plants found in the county : with sketches of its physical geography and climate and of the progress of Middlesex Botany during the last three centuries*. R. Hardwicke, London.

John Ray

Birkhead, T., (ed.), (2016), *Virtuoso by Nature: the scientific worlds of Francis Willughby FRS (1635–1672).* Brill, Leiden.

Birkhead, T. (2018), *The wonderful Mr Willughby: the first true ornithologist.* Bloomsbury Publishing, London.

Lankester, E., (ed.), (1848), *The Correspondence of John Ray.* Ray Society, London.

Oswald, P.H. and Preston, C.D., (eds.), (2011), *John Ray's Cambridge Catalogue (1660).* Ray Society, London.

Raven, C.E. (1950), *John Ray Naturalist: his life and works*, 2nd edn. Cambridge University Press, Cambridge.

Reveal, J.L., Broome, C.R., Brown, M.L. and Frick, G.F. (1987), The identification of pre-1753 polynomials and collections of vascular plants from the British colony of Maryland. *Huntia,* 7: 91–208.

Roos, A.M., (ed.), (2015), *The Correspondence of Dr. Martin Lister (1639–1712). Vol. One: 1662–1677.* Brill, Leiden.

Richard Richardson

Edgington, J. (2016). Natural history books in the library of Dr Richard Richardson, *Archives of Natural History* 43 (1): 58.

Loudon, J. C. (ed.), (1828), Some account of Richard Richardson esq. *Gardener's Magazine*, 3: 127–8.

Pulteney, R. (1790). *Historical and biographical sketches of the progress of botany in England*, Vol. 2. Printed for T. Cadell, London.

Scott, E.J.L. (1904), Index to the Sloane manuscripts in the British Museum. British Museum, London.

Smith, J. E. (ed.), (1821), *A Selection of the Correspondence of Linnaeus, and Other Naturalists, from the Original Manuscripts*, 2 vols. Longman, Hurst, Rees, Orme, and Brown, London.

Turner, D. (1835), *Extracts from the Literary and Scientific Correspondence of Richard Richardson, M.D., F.R.S., of Bierley, Yorkshire*. Printed by Charles Sloman, King Street, Yarmouth.

Joan Salvador i Riera

Bolòs, A. (1947), Plantas montserratinas de Juan Salvador. *Collectanea Botanica,* 1: 323–329.

Camarasa, J.M. and Ibáñez, N. (2007), Joan Salvador and James Petiver: a scientific correspondence (1706–1714) in time of war, *Archives of Natural History* 34(1): 140–173.

Camarasa, J.M. and Ibáñez, N. (2012), Joan Salvador and James Petiver: the last years (1715–1718) of their scientific correspondence. *Archives of Natural History,* 39(2): 191–216.

Dandy, J.E. (1958), *The Sloane Herbarium.* an *annotated list of the Horti Sicci composing it: with biographical accounts of the principal contributors*. British Museum (Natural History), London.

Ibáñez, N. (2006), *Estudis sobre cinc herbaris històrics de l'Institut Botànic de Barcelona*. PhD thesis, University of Barcelona, Spain.

Ibáñez, N., Camarasa, J.M. and Garcia-Franquesa, E. (Coords.) (2019*), El Gabinet Salvador. Un tresor científic recuperat*. Manuals del Museu, 2. Museu de Ciències Naturals de Barcelona, Barcelona. https://doi.org/10.32800/manuals.2019.0002

Ibáñez, N., Montserrat, J.M., Soriano, I. and Camarasa, J. M. (2006), Plant material exchanged between James Petiver (c.1663–1718) and Joan Salvador i Riera (1683–1725). I. The Balearic plants conserved in BC-Salvador and BM-Sloane Herbaria. *Notes and Records of the Royal Society of London,* 60: 241–248.

Pardo-Tomàs, J. (2014), *Salvadoriana. The Cabinet of Curiosities of Barcelona*. Gràfiques Alpres, Barcelona.

Petiver, J. (1695–1703), Musei Petiveriani Centuria. Samuel Smith & Benjamin Walford, London.

Petiver, J. (1716), Plants gathered on Mount Serrato lately sent me from Mr. John Salvadore at Barcelona. *Petiveriana seu Naturae Collectanea,* 2: 8.

Petiver, J. (1717), Plantae Baleares. Plants found in the islands Majorca and Minorca, by the curious Mr. John Salvadore, apothecary at Barcelona, and Mr. George Bouchere, surgeon at Port Mahone. *Petiveriana seu Naturae Collectanea,* 3: 9.

Richter, H.E. (2003), *Codex Botanicus Linneanus*. Regnum vegetabile, v.140. Gantner, Liechtenstein.

Rosselló, J.A and Sáez, Ll. (2000), Index Balearicum: An annotated checklist of the vascular plants described from the Balearic Islands. *Collectanea Botanica*, 25(1): 3–203.

William Sherard

Clokie, H.M. (1964), *An Account of the Herbaria of the Department of Botany in the University of Oxford*. Clarendon Press, Oxford.

Delbourgo, J. (2017), *Collecting the World: the Life and Curiosity of Hans Sloane*. Allen Lane, London.

Harris, S.A. (2015). William Sherard: his herbarium and his *Pinax*. *Oxford Plant Systematics,* 21:13–15.

Jackson, B.D. (1874), A sketch of the life of William Sherard. *Journal of Botany, British and Foreign*, 12:129–138.

Riley, M. (2011), Procurers of plants and encouragers of gardening: William and James Sherard and Charles du Bois, case studies in late seventeenth- and early eighteenth-century botanical and horticultural patronage. Ph.D. Thesis, University of Buckingham.

Smith, J.E. (1816), Sherard, William, in: *Dr Rees's New Cyclopaedia*, Vol. 32, part 2. Longman, Hurst, Rees, Orme and Brown, London.

Turner, D. (1835), *Extracts from the Literary and Scientific Correspondence of Richard Richardson, M.D., F.R.S., of Bierley, Yorkshire*. Printed by Charles Sloman, King Street, Yarmouth.

Sir Hans Sloane

Carney, J. (2010), *In the Shadow of Slavery: Africa's Botanical Legacy in the Atlantic World*. University of California Press, Berkley.

Delbourgo, J. (2017), *Collecting the World: The Life and Curiosity of Hans Sloane.* Allen Lane, London.

Hayden, C. (2003), *When Nature Goes Public: The Making and Unmaking of Bioprospecting in Mexico*. Princeton University Press, Princeton.

Schiebinger, L. (2004), *Plants and Empire: Colonial Bioprospecting in the Atlantic World*, Harvard University Press, Cambridge, MA.

Mary Somerset, Duchess of Beaufort

Chambers, D. (1997), Storys of plants: the assembling of Mary Capel Somerset's botanical collections at Badminton. *Journal of the History of Collection,* 9 (1): 49–60.

Kell, P.E. (2004), Somerset, Mary, Duchess of Beaufort (*bap.*1630, *d.*1715). *Oxford Dictionary of National Biography*, online edition.

Laird, M. (2015), Nursing pretty monsters – the Duchess of Beaufort's Florilegium and herbarium and the art of Kickius. In: M. Laird, *A Natural History of Gardening 1650–1800*. Yale University Press, London, pp 62–123.

McClain, M. (2001), *Beaufort: the Duke and his Duchess 1657-1715*. Yale University Press, London.

Munroe, J. (2011), 'My innocent diversion of gardening': Mary Somerset's plants, *Renaissance Studies* 25 (1): 111–123.

Wynne Smith, L. (2012), Sloane as a friend and physician of the family. In: A. Walker et al (eds.), *From Books to Bezoars*, The British Library, London, pp.48–56.

Bruno Tozzi

Pichi Sermolli, R.E.G. (1999), Contributo alla storia della botanica in Toscana. I precursori dell'esplorazione floristica delle Alpi Apuane. *Museologia Scientifica*, 15 (2): 1–289.

Robert Uvedale

Petiver, J. (1713), Botanicum Hortense. III. Giving an Account of Divers Rare Plants, Observed This Summer, A.D. 1713, in Several Curious Gardens about London, and Particularly the Society of Apothecaries Physick Garden at Chelsea, by James Petiver, F.R.S, *Philosophical Transactions of the Royal Society of London* 28 (337): 177–221.

Farrington, A. (1999), *A Biographical Index of East India Company Maritime Service Officers 1600–1834*. The British Library, London.

Turner, D. (1835), *Extracts from the Literary and Scientific Correspondence of Richard Richardson, M.D., F.R.S., of Bierley, Yorkshire*. Printed by Charles Sloman, King Street, Yarmouth.

Pulteney, R. (1790). *Historical and biographical sketches of the progress of botany in England*, Vol. 2. Printed for T. Cadell, London.

Sloane, H. (1707–1725), *A Voyage to the Islands Madera, Barbados, Nieves, S. Christophers and Jamaica, with the Natural History of the Herbs and Trees, Four-footed Beasts, Fishes, Birds, Insects, Reptiles, &c...* London.

About the authors

Esperanza Maribel Agoo is a Professor of Biology at the De La Salle University, Manila. She manages the DLSU Herbarium and Seed Bank. Her main research interests are systematics, conservation, ecology, and ethnobotany of Philippine seed plants. She is also interested in historical botany and the plants collected by Father Kamel in the Philippines.

Mark Carine is Principal Curator in Charge of the botanical collections at the Natural History Museum. He is interested in the diversity and evolution of oceanic island floras; work that has involved research on early records of island plants in the Sloane Herbarium and the use of those collections to help understand human impact on island floras.

Richard Coulton is Senior Lecturer in the Department of English at Queen Mary University of London. He researches intersections between the histories of horticulture, science, landscape, and empire in the eighteenth century. Recent work in collaboration with colleagues at the Natural History Museum has focused on James Petiver's global collecting network, with particular reference to the East India Company world.

James Delbourgo is Professor of History at Rutgers University, where he teaches the history of science, collecting and museums, and author of *Collecting the World: The Life and Curiosity of Hans Sloane* (Allen Lane, 2017). His current research projects include the history of collecting, global histories of science and the history of swimming and underwater exploration.

Stephen Harris is Curator of Oxford University Herbaria. He is interested in the evolutionary consequences of plant movement, early modern botany and the use of herbaria as resources for answering questions in modern plant sciences.

Robert Huxley was Head of Botanical Collections and is now a Scientific Associate at the Natural History Museum. He writes on the history of natural history and is interested in training and development in the strategic management of natural science collections. He is also an Honorary Curator at World Museum Liverpool.

Neus Ibáñez is curator of the herbarium at the Institut Botànic de Barcelona (BC). She holds a PhD in Biology, specialized in botany and specifically in the history of botany and the study of historical herbaria. She was curator of the exhibition "Salvadoriana", exhibited at the Institut Botànic de Barcelona (2014–2016).

Charles Jarvis is a Scientific Associate at the Natural History Museum with interests in the history of botanical exploration in the late 17th and early 18th centuries. His current research focuses on the rôle played by the Scottish surgeon James Cuninghame, a prolific early collector of plants in China whose specimens feature prominently in the Sloane Herbarium.

Sebestian Kroupa is a Postdoctoral Research Fellow at King's College London. A historian of early modern natural history and medicine, his recently completed PhD at the University of Cambridge traced the communication of natural and medical knowledge between the Philippines and Europe in the early eighteenth century, drawing on the plant specimens, texts and images of Georg Joseph Kamel.

Sachiko Kusukawa is Professor of History of Science at the University of Cambridge and Fellow of Trinity College, Cambridge. She is interested in early modern collections that include natural historical specimens as well as images, including those of William Courten and Hans Sloane.

Domingo A. Madulid is Professorial Lecturer, Biology, at De La Salle University, Manila and retired Chief Botanist, National Museum. His main interests are historical botany, floristic inventory and taxonomy of Philippine plants. He has published on Father Manuel Blanco and the naturalists of the Malaspina expedition. His other interest include early naturalists in the Philippines including Father Georg Joseph Kamel.

Kathleen Murphy is Professor and Chair of the History Department at California State Polytechnic University in San Luis Obispo, California. She is interested in the history of science of the eighteenth-century British Atlantic World and, especially, the entangled histories of the slave trade and natural history. Her research relies on the many records associated with Hans Sloane's collections, including his herbarium.

Victoria Pickering is a Fellow at Dumbarton Oaks working as part of the Plant Humanities Initiative. Focusing on the seventeenth and eighteenth centuries, Victoria is interested in the collection, movement, management, and use of natural history objects, paying particular interest to the context of the people and relationships who enabled this global movement of knowledge.

Laia Portet i Codina has a History degree from the Universitat Autònoma de Barcelona, a Masters in World History from the Umiversitat Pompeu Fabra and a Masters in History and Culture of Food from the University of Barcelona. She is a PhD candidate at Cambridge University with research interests revolving around knowledge, commerce and science in the early modern world.

Edwin Rose is completing a PhD at the University of Cambridge and from October 2020 will be Munby Fellow in Bibliography at Cambridge University Library and a research fellow at Darwin College, Cambridge. He has interests in the history of the life and earth sciences and the history of the book from the late seventeenth to late nineteenth centuries.

Ranee Prakash is a Senior Curator in the General Herbarium at the Natural History Museum with interests in herbarium collections, tropical botany and ethnobotany (especially medicinal plants). She is currently involved in a research project dealing with the collections of Samuel Browne from late 17th century peninsular India to investigate plant use through time and space across different cultures.

Chris Preston worked before retirement at the Biological Records Centre, compiling atlases of vascular plants and bryophytes in Britain and Ireland, a botanical equivalent of the lexicographical harmless drudge. In retirement he is continuing to pursue his interests in the history of British botany and the bryophytes of Cambridgeshire, and has begun to study plant parasitic microfungi.

Fred Rumsey is Senior Curator in Charge of the British, Fern and Historical collections at the Natural History Museum. He is interested in the British flora, the history of its discovery and its conservation; work that has involved research on early records in the Sloane Herbarium and the use of those collections to help understand status and inform conservation prioritisation.

Suzanne Ryder is Senior Curator in Charge of the Hymenoptera and Historical Entomology collections at the Natural History Museum. Suzanne is also the honorary curator of Entomology for the Linnean Society. She is interested in historical collections and their conservation and her current area of research includes Sir Josephs Banks's insect collection.

Adriano Soldano is mainly interested in early botanical research in Italy in order to produce a new chronology of the vascular plants of the country; the work is combined with manuscript consultation. He is also interested in taxonomy, nomenclature and floristic exploration.

Mark Spencer is a consultant botanist, honorary curator of the Linnean Society of London's herbariums and a Scientific Associate at the Natural History Museum. He is particularly interested in 17–18th century English botanical exploration and the introduction of 'exotick' plants into European horticulture during that period.

Tod Stuessy is Professor Emeritus at the University of Vienna and The Ohio State University. His research focuses on the evolution and biogeography of endemic flowering plants of oceanic islands, particularly in the Juan Fernández Archipelago. Other interests include concepts and methods of biological classification, and phylogeny and speciation of Latin American Asteraceae.

Jacek Wajer is curator in charge of Section II of the General Herbarium at the Natural History Museum. He is interested in the influence of horticulture on the development of the museum's botanical collections. His research is currently focused on the origin of the specimens collected by Philip Miller and other important 18th century gardeners in the Sloane Herbarium.

Anna Winterbottom is a Research Associate at McGill University. She works on the history of science, medicine, and the environment, with a particular focus on the Indian Ocean region. Her writing includes *Hybrid Knowledge in the Early East India Company World* (Palgrave, 2016). She is currently researching the early modern circulation of materia medica.

Index

Picture Credits

p.2-3 H.S. 311; f. 6; p.11 H.S. 318; f. 75;p.9 Ray (1686) Historia Plantarum 1: 126-127. (NHM L&A); p.13 ©Malcolm Penn; p.14 H.S. 3; f. 35; p.15 H.S. 44; f. 33; p.16 detail from H.S. 92 f. 7; p.18 H.S. 319; f. 87; p.21 H.S. 60 & 82; f. 49; p.23 H.S. 59; f. 48; p.24 H.S. 59; f. 51; p.25 H.S. 54; f.67; p.27 H.S. 150; f. 60; p.29 H.S. 157; f. 8; p.30 H.S. 156; f. 244; p.31 Petiver (1702) Gazophylacii naturae et artis, t. 7. (NHM L&A); p.33 Micheli (1729) Nova Plantarum Genera, t. 91 (NHM L&A); p.34 Micheli (1729) *Nova Plantarum Genera*, t. 83 (NHM L&A); p.35 H.S. 149; f. 24; p.36 Micheli (1729) *Nova Plantarum Genera*, frontispiece (NHM L&A); p.37 ©Ralf Roletschek assumed (based on copyright claims). / CC BY-SA (http://creativecommons.org/licenses/by-sa/3.0/); p.39, 42, 43, 160, 161 © Oxford University Herbaria; p.40 (top) ©Department of Plant Sciences, University of Oxford; p.40 (bottom) ©Sherardian Library of Plant Taxonomy, Bodleian Library;p.45 H.S. 149; f. 112; p.46 Plukenet (1700) Almagesti botanici mantissa, t. 333 (NHM L&A); p.47 H.S. 92 f. 7; p.48 Plukenet Entomological Collection, f. 104; p.49 Plukenet (1691) *Phytographia* I: frontispiece (NHM L&A); p.50, 106 ©Wellcome Collection; p.51 H.S. 37; f. 70; p.53 Ray (1704) Historia plantarum 3: 594 (NHM L&A); p.54 H.S. 37; f. 107; p.56 H.S. 167; f. 97; p.58, 197 ©Mark Carine; p.60 ©National Portrait Gallery,London NPG D30510; p.62 H.S. 142; f. 56; p.63 H.S. 131; f. 16 & Sloane Vegetable substances no. 5055 (NHM L&A); p.64 H.S. 131; f. 52; p.65 H.S. 142; f. 67; p.67 H.S. 312; f. 72; p.69 H.S. 303; f. 81; p.70 H.S. 314; f. 33; p.71 H.S. 312; f. 81; p.73 Martyn (1728) Historia plantarum rariorum, Centuria 1, Decas 1-5, t. 21 (NHM L&A); p.74 Miller (1731) *The Gardeners Dictionary*, 1st Edition. (NHM L&A); p.76 H.S. 229; f. 1; p.77 Martyn (1728) *Historia plantarum rariorum*, Centuria 1, Decas 1-5, t. 41 (NHM L&A); p.79 H.S. 167; f. 329; p.80 H.S. 167; f. 425; p.81 H.S. 167; f. 411; p.82 H.S. 167; f. 45; p.83 H.S. 167; f. 97; p.85 detail from H.S. 115: f. 18; p.86, 95 ©Fred Rumsey; p.87 H.S. 85*; f. 82; p.88 Sloane (1725), *A voyage to the islands Madeira, Barbados, Nieves, S. Christophers and Jamaica*, vol 2: t 173 (NHM L&A); p.90 H.S. 125: f. 32 (left) and H.S. 125: f. 5 (right); p.91 H.S. 115: f. 18; p.93 H.S. 124: f. 30; p.96 H.S. 152; f. 102; p.97 H.S. 152; f. 156; p.98 H.S. 54; f.61; p.100 Sloane's Vegetable Substances catalogue, vol 3; f. 932 (NHM L&A); p.101 H.S. 302; f. 66; p.102 Ray (1724) *Synopsis Methodica Stirpium Britannicarum*, t. 3 (NHM L&A); p.103 H.S. 79; f. 130; p.104 H.S. 145; f. 2; p.107 Petiver (1767) *Gazophylacii Naturae & Artis*, t. 129 (NHM L&A); p.108 © Angelo Mazzoni; p.109 H.S. 149, f. 86; p.110 H.S. 149, f. 106; p.112, 113, 114, 115 ©Consorci del Museu de Ciències Naturals de Barcelona. Photography: Jordi Vidal F; p.117 H.S. 166; f. 152; p.118 detail of H.S. 292; f. 14; p.120 Sloane (1707), *A voyage to the islands Madeira, Barbados, Nieves, S. Christophers and Jamaica*, vol. 1. (NHM L&A); p.121 H.S. 316; f. 40; p.123 Bosman (1814) *A new and accurate description of the Coast of Guinea divided into the Gold, the Slave and the ivory Coasts : written originally in Dutch ... and now faithfully done into English, &c.* (NHM L&A); p.124 H.S. 154; F. 66; p.125 Petiver (1711) Gazophylacii Naturae & Artis, t. 69 (NHM L&A); p.127 H.S. 155; f.161; p.130 Sloane (1707) *A voyage to the islands Madeira, Barbados, Nieves, S. Christophers and Jamaica*, vol. 1. (NHM L&A); p.131 Sloane (1725), *A voyage to the islands Madeira, Barbados, Nieves, S. Christophers and Jamaica*, vol 2: t. 160 (NHM L&A); p.132 H.S. 5; f. 59; p.133 H.S. 2; f. 2; p.134 ©The John Bartram Association, Bartram's Garden, Philadelphia; p.135 H.S. 145; f. 59; p.137 H.S. 242 title page; p.138 H.S. 145; f.48; p.141 H.S. 292; f. 14; p.142 H.S. 292; f. 20; p.143 Houstoun (1730?) *Drawings of plants from Central America chiefly from Vera Cruz* (manuscript; NHM L&A); p.144 H.S. 116; f. 11; p.145 H.S. 116; f. 17; p.147 H.S. 145; f. 8; p.149 H.S. 74; f. 45; p.150 Solander slip catalogue XIII: 269 (NHM L&A); p.156 H.S. 212; f. 56; p.159, 153 ©Sherardian Library of Plant Taxonomy, Bodleian Library; p.161 H.S. 334; f. 92 (left) and H.S. 332; f. 7 (right); p.163 H.S. 332; f. 9; p.164 detail of H.S. 211; f. 27; p.167 Sloane Vegetable Substances Collection no. 8907; p.169 H.S. 105; f. 10; p.171 H.S. 330; f. 6; p.173 Herbarium of Samuel Browne, vol 1; f. 2; p.175 H.S. 165; f. 385; p.177 H.S. 231; f16; p178 H.S. 165; f.389; p.179 © The British Library Board, Sloane MS 4080, f. 122; p.181 H.S. 331; f. 113; p.182 H.S. 94; f. 157 (left); Plukenet (1705) Amaltheum Botanicum, t. 371, f. 1 (NHM L&A) (right); p.183 ©Image by Charles E. Jarvis; p.184 Plukenet (1705) *Amaltheum Botanicum,* t. 390, f. 2. (NHM L&A); p.185 H.S. 94; f. 192; p.187 H.S. 211; f. 27; p.188 Kaempfer (1727) *The History of Japan*, tab 38 (NHM L&A); p.189 Kaempfer (1727) *The History of Japan*, t. 39 (NHM L&A); p.190 H.S. 211; f. 103 (top); Kaempfer (1712) *Amoeniatum exoticarum politico-physico-medicarum fasciculi V*, opposite p. 812. (NHM L&A); p.191 Kaempfer (1727) *The History of Japan*, tab 8 (NHM L&A); p.193 H.S. 8; f. 19; p.194 H.S. 8; f. 100; p.195 Sloane (1693) *Philosophical Transactions of the Royal Society*, 17: 204; p.197 ©Juliet Brodie

Conventions used:

The 'H.S.' number is the Hortus Siccus number in the Sloane Herbarium.

Sources other than the botanical collections at Natural History Museum, London Library & Archives (NHM L&A) are given in parentheses after details of the image.